Praise for

EAT UP

A smart, comprehensive guide for ambitious beginners written by a woman with extensive expertise in her rooftop field. If we're going to see gardening spread atop our skylines, it'll be led by farmers with this book in their backpack as they navigate the intersection of people, plants and policy. This guide suits all cities, roofs and styles of farming: once read, you can no longer ignore the possibilities above us. Two green thumbs up!

— Annie Novak, farmer and cofounder, Eagle Street Rooftop Farm

Equitable access to food is a global concern. Landscape architect Mandel has given us an important book that provides a glimpse into the burgeoning world of rooftop agriculture to meet this formidable challenge. *EAT UP* is an engaging and informative resource loaded with empowering information for individuals and communities on the possibilities for creating our own multi-scalar rooftop food systems.

— Lucinda Sanders, FASLA, CEO, Olin

Lauren Mandel offers a practical introduction to the exciting potential as well as the challenges associated with rooftop agriculture. *EAT UP* is a must-read for designers, urban planners and agriculturalists who are thinking seriously about how our cities can accommodate the pressing needs of the 21st century. This book offers useful case studies of successful projects which serve a wide range of values, including building community, promoting social justice and bolstering public health.

— Charlie Miller, president, Roofmeadow

EAT UP is an inspirational look at the future of food gardening in cities. This book will expand your notions of where gardening is possible and just how much food you really can grow right at home.

— David Greenberg, Executive Director, Growing Gardens

Developing a resilient, diversified food system presents an amazing opportunity to revitalize communities while simultaneously bringing social, economic and environmental benefits. Rooftop agriculture undoubtedly presents an underutilized piece of this puzzle. Anyone interested in bringing the benefits of a localized food system to cities will gain from *EAT UP*'s clear and concise information on how to capture the relatively untapped potential presented by urban rooftops.

—Beth McKellips, Director, Cornell Cooperative Extension
Agricultural Economic Development

Rooftops may be America's greatest unused resource, and Lauren Mandel's book gives an exciting, but realistic assessment of the opportunities and challenges involved with turning these forgotten spaces into productive sources of food. We want to empower Americans from all backgrounds to get outside and build community. What better way than if everyone could traverse the skyline?

— Stacy Bare, director, Mission Outdoors, Sierra Club

EAT UP

The Inside Scoop on
ROOFTOP
AGRICULTURE

LAUREN MANDEL, MLA

new society
PUBLISHERS

Cover design by Diane McIntosh.
City image © iStock (albertc111); Vegetable images: © iStock (Keith Bishop)
All interior photographs © Lauren Mandel unless otherwise indicated.
Author photograph by Geoffrey Goldberg Photography.

Printed in Canada. First printing April 2013.
Paperback ISBN: 978-0-86571-735-0
eISBN: 978-1-55092-530-2

Inquiries regarding requests to reprint all or part of *EAT UP*
should be addressed to New Society Publishers at the address below.

To order directly from the publishers, please call toll-free (North America) 1-800-567-6772,
or order online at www.newsociety.com

Any other inquiries can be directed by mail to:

New Society Publishers
P.O. Box 189, Gabriola Island, BC V0R 1X0, Canada
(250) 247-9737

New Society Publishers' mission is to publish books that contribute in fundamental ways to building an ecologically sustainable and just society, and to do so with the least possible impact on the environment, in a manner that models this vision. We are committed to doing this not just through education, but through action. The interior pages of our bound books are printed on Forest Stewardship Council®-registered acid-free paper that is **100% post-consumer recycled** (100% old growth forest-free), processed chlorine free, and printed with vegetable-based, low-VOC inks, with covers produced using FSC®-registered stock. New Society also works to reduce its carbon footprint, and purchases carbon offsets based on an annual audit to ensure a carbon neutral footprint. For further information, or to browse our full list of books and purchase securely, visit our website at: **www.newsociety.com**

Library and Archives Canada Cataloguing in Publication

Mandel, Lauren
    Eat up : the inside scoop on rooftop agriculture / Lauren Mandel.
Includes bibliographical references and index.
ISBN 978-0-86571-735-0

    1. Roof gardening. 2. Urban gardening. 3. Urban agriculture. 4. Local foods. I. Title.
SB419.5.M35 2013   635.9'671   C2013-900850-0

For Leah (1980-2009)
and her pursuit of food justice.

Contents

Acknowledgments

First and foremost, thank you to the countless rooftop farmers, gardeners, educators, volunteers, CEOs, chefs and other industry leaders whose willingness to collaborate made this project a reality. Thank you also to the editorial, production and marketing teams at New Society Publishers, in particular, Ingrid Witvoet. A very special thank you to Caroline Dlin and Eric Cohen whose guidance and generosity helped shepherd me through the manuscript development and contract negotiation processes. Thank you also to my coworkers at Roofmeadow whose professional support made this book possible. Thank you to my advisors at the University of Pennsylvania — Karen M'Closkey, Charlie Miller, Domenic Vitiello and Anita Mukherjee — who provided valuable feedback during *EAT UP*'s fledgling stage.

Thank you to the photographers who contributed their work to this collective effort: Geoffrey Goldberg, Michael Mandel, Jake Stein Greenberg, Ari Burling, John Q. Porter, Allen Ying, Trenton Barnes, Karen Jacobson, Luke Mitchell and Anastasia Cole Plakias.

Thank you also to Michael Miller, Karen Lutsky, Alexa Bosse, Aditi Sen, Harley Cooper, Sue Van Hook, David Gouverneur, Andee Mazzocco, Nic Esposito, Bill Shick, Geoffrey Dlin, Bret Betnar, Jessica Henson, Gavin MacIntyre, Emily Mastellone-Snyder and Aaron Kelley. Lastly, thank you to my family — Paula Mandel, Richard Mandel and Michael Mandel — for their relentless support, and to my dog Sūka, to whom I owe many walks.

Eagle Street Rooftop Farm, NY
PHOTO BY ISAIAH KING

Intent

From warehouse roofs in Brooklyn to the tops of churches in San Francisco, urbanites across North America are rolling up their sleeves, reaching deeply into the soil and taking back our food system. Rooftop agriculture is real. It's happening. Every day the movement's energy infiltrates further into schools, restaurants and office buildings — captivating people's stomachs and their souls. The power of these spaces can be seen and heard and tasted as skyline farmers deliver fresh kale and strawberries to our tables. Their power can be felt as rooftop gardeners spread the word of community initiative and healthy eating. It's happening.

In 2009, I began exploring the viability of rooftop agriculture — an unconventional method of food production at the time.

In a few short years, a constellation of new farms and gardens across our city skylines reveals the industry's extreme growth, and unparalleled potential for expansion. This movement is ripe for harvest. *EAT UP* provides the tools you need to turn your dreams of rooftop farms and gardens into actual spaces that feed people. With collective insight from the leading rooftop farmers, gardeners, educators, volunteers, CEOs, chefs, designers and green roof professionals, *EAT UP* is the most comprehensive resource to date on rooftop agriculture. Don't just read the book; use it to cultivate your own rooftop expertise, and feed your friends along the way.

Now climb up there and get started!

1 | The Backdrop

Rooftop greens basking in the sun,
Uncommon Ground, IL

what is rooftop agriculture?

Soaring food prices and obesity rates increasingly prompt North Americans to grow their own fruits and vegetables close to home. In cities, forgotten parcels such as vacant lots, sidewalk strips, and park fragments historically served as prime poaching grounds for urban farmers and gardeners to plant their seeds. During recent years, however, land insecurity and contaminated soils demand creative solutions that allow urban agriculture to creep up walls and balconies, and onto rooftops. Broadly speaking, rooftop agriculture is the cultivation of plants, animals and fungi *on rooftops* for the purpose of human use and consumption. This includes foodstuffs, fibers, animal products and medicinal plants. The hunger for local food has reached new heights, and you truly can't get more local than your own roof!

urban agricultural niche

Rooftop agriculture is one cog in the greater urban food system. The practice should not be viewed as a cure-all for hunger, nor should the assumption be reached that it will dominate food production in all cities. Rooftop agriculture works in concert with community gardens, farmers' markets, grocery stores and, of course, rural agriculture to feed hungry cities. A key principle of ecology states that diversity in any system breeds resilience. If one strand of the web fails, the others will hold the web together. Food systems are no different. Rooftop agriculture is powerful in its ability to enhance the diversity, and therefore resilience, of the greater urban food system. Farmers and gardeners pursuing all types of urban agriculture, from planting sunflowers along abandoned railways to raising fish in basements, have the potential to learn from one another. Rooftop agriculture similarly absorbs lessons from other forms of urban production, and in turn contributes to the collective knowledge. There's always more to learn.

gardening vs. farming

The boundary between gardening and farming is a blurry one. Practitioners, academics and even policy makers qualify the distinction in varying ways, and no one can seem to agree upon a universal definition. One common and relatively compelling opinion describes gardening as the production of agricultural products for self-consumption, charity or gifting. Farming is often defined as the production of these same goods in exchange for money. *EAT UP* embraces this distinction and highlights inspirational gardens *and* farms on rooftops around North America. The book also explores the rooftop agricultural industry, as this scale of production encompasses both gardening and farming which involves rooftop farms and gardens with multiple locations.

> Rooftop agriculture is powerful in its ability to enhance the diversity and resilience of the greater urban food system.

> The hunger for local food has reached new heights.

who's doing it?

As rooftop farms and vegetable gardens sprout up in cities across North America, restaurant patrons, community groups, individuals and families get to savor the bounty. But who's actually up on the roof growing all this food? Lots of people — that's who! People of all different ages and ethnicities, with varied skill sets, and dozens of reasons for growing food.

In North America, rooftop farmers tend to be between the ages of 22 and 55, with men and women equally engaged. Most skyline farmers migrate to rooftops from ground-level farms, some in urban areas, others in more traditional rural settings. Farmers who land these highly prized rooftop positions are generally very knowledgeable about their agrarian genre — whether it be row farming, beekeeping, hydroponics or some other form of production. It's rare that a newbie finds herself in charge of much on the skyline, as these farms can require large initial investments, which leaves little room for error with day-to-day operations. Less experienced apprentices, interns and volunteers often assist rooftop farmers, as do teams of directors, volunteer coordinators, marketing personnel, publicity coordinators and technical specialists. Other professionals that are critical during the rooftop farm's design and construction may include a landscape architect, green roof consultant, structural engineer, mechanical/electrical/plumbing (MEP) engineer, architect, waterproofing membrane provider, greenhouse designer, hydroponic system designer and construction contractors from various trades. It takes a village.

Rooftop gardening, on the other hand, attracts enthusiasts with all levels of experience. Novice gardeners may enjoy planting rooftop containers with herbs, while a master gardener may forge an entire community garden by himself. Kids play an important role in rooftop gardening as well. From school and after-school gardens, to family plots, to community gardens, kids spark enthusiasm and soak up knowledge about rooftop gardening. They even teach their friends and parents! Other parties that may be involved during the inception and construction of a rooftop garden include a structural engineer, architect, carpenter, plumber, and possibly an electrician.

Whether well-seasoned or completely green, urbanites from all backgrounds increasingly seek out rooftop farming and gardening opportunities. As the trend continues, more and more individuals, communities and entrepreneurs will look toward the roof for a food solution.

This book highlights North American rooftop agriculture professionals and enthusiasts from Canada, Lebanon, India, France, Sweden, Australia and all corners of the US.

Volunteer harvesting herbs,
Eagle Street Rooftop Farm, NY

a brief history

urban agriculture

The earliest record of food production within cities dates back to **3100 BCE**, when home vegetable gardens were commonplace in China.[1] Some historians believe that in addition to providing food, these gardens were built as outlets for recycling organic waste generated by the household. Early Latin American communities may also have practiced urban agriculture, predominantly as a means of promoting food security within cities. Throughout the rest of the developed world, urban populations commonly relied upon locally produced food until the Industrial Revolution of the late eighteenth and early nineteenth centuries. This age of mechanization and industry changed everything for food production, as factories and urban development began taking the place of agricultural plots. The result? Agriculture moved outward from city centers to form rural "belts." Urban agricultural planner Jac Smit suggests that this agricultural migration away from cities directly leads to a need for food importation *into* cities.[2]

Since the Industrial Revolution agriculture has periodically reentered cities, most notably, perhaps, with the planting of Victory Gardens during World Wars I and II to supplement the public food supply. Currently, the urban agriculture trend is gaining steam, as cities around the globe recognize the benefits of food localization. The Netherlands, for example, produced 33% of its domestic agricultural needs within urban areas in 2000,[3] and the Institute for Food and Development estimated that in 1999, **14% of the global food supply was produced in cities**.[4] As cities densify, expand and multiply in number, agriculture's reoccupation of urban space will continue to spread.

rooftop agriculture

As with urban agriculture, rooftop agriculture possesses a lengthy history, which likely dates back to **600 BCE Babylon** (present day Iraq). Geoff Wilson, a noted authority on urban agriculture, believes that the Hanging Gardens of Babylon were likely "the world's first rooftop farming project."[5] Archeological evidence suggests that these terraced roof gardens were used to produce fruit, vegetables and possibly even fish! During the 1500s, the Aztecs may have also built sophisticated rooftop farms, which incorporated waste management strategies. The current trend in rooftop farming is predominantly fueled by Canada, the US and Singapore. Other countries around the world, such as Australia, Senegal, Russia, Italy, India, Egypt and Hong Kong, are also beginning to explore rooftop food production within cities. The agricultural methods used within each country reflect local climatic, culture, socioeconomics and building characteristics.

Rooftop agriculture's lengthy history ... likely dates back to 600 BCE Babylon

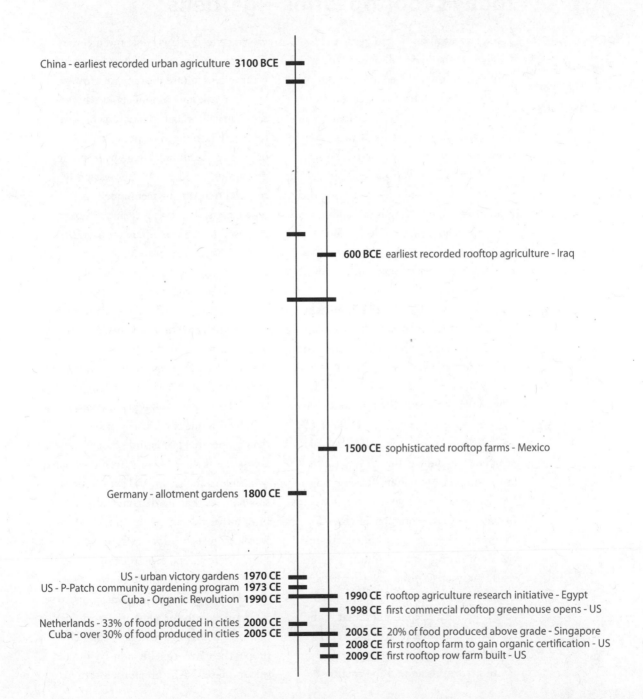

China - earliest recorded urban agriculture **3100 BCE**

600 BCE earliest recorded rooftop agriculture - Iraq

1500 CE sophisticated rooftop farms - Mexico

Germany - allotment gardens **1800 CE**

US - urban victory gardens **1970 CE**
US - P-Patch community gardening program **1973 CE**
Cuba - Organic Revolution **1990 CE**
1990 CE rooftop agriculture research initiative - Egypt
1998 CE first commercial rooftop greenhouse opens - US
Netherlands - 33% of food produced in cities **2000 CE**
Cuba - over 30% of food produced in cities **2005 CE**
2005 CE 20% of food produced above grade - Singapore
2008 CE first rooftop farm to gain organic certification - US
2009 CE first rooftop row farm built - US

today's rooftop farms + gardens

Rooftop farms and gardens address more than just subsistence.

Furthering the rooftop agriculture movement as a whole benefits everyone.

Today's rooftop farms and gardens address more than just subsistence. They are built to foster healthy eating, community building, stormwater management, business development and the occupation of underutilized space. Many of these skyline gems fulfill multiple goals at once, such as an educational rooftop garden that teaches kids about nutrition while supplying the cafeteria and improving local food security.

From schools to roof decks, churches to restaurants, apartment buildings to warehouses, rooftop farms and gardens sprout up in all types of neighborhoods and engage people from all walks of life. They occupy industrial zones, high-rent districts and even financial centers. Rooftop farms and gardens appear in various shapes and sizes, and consequently produce a wide range of yields. Generalizing about such dynamic spaces is tricky, and so **EAT UP** groups these farms and gardens by scale of operation and whether or not the sites are commercial in nature. The resulting typologies are therefore: Rooftop Gardens (small-scale, non-commercial), Rooftop Farms (medium-scale, commercial) and the Rooftop Agriculture Industry (large-scale, commercial). The following pages discuss these typologies in more detail, so for now, just keep in mind that rooftop farms and gardens are diverse places intended for a variety of purposes.

In my travels around the country visiting sites for this book and meeting the men, women and children behind the rooftop agriculture movement, growers often ask my opinion about whether their agricultural production technique (e.g., row farming) is superior to "the other guy's technique" (e.g., hydroponic greenhouse production). I consistently respond that no one technique is better than any other; they're just different, and used to achieve different goals. This message is important to remember, as each type of operation fulfills specific objectives, and benefits specific types of people, but furthering the rooftop agriculture movement as a whole benefits everyone.

With that said, the existing North American rooftop agriculture movement, as well as the industry, is in a fledgling state. The current "boutique" industry, if you will, predominantly consists of independently operated farms and gardens dotted across the skyline. No organizational entity coordinates the effort, and some cities possess policies that are detrimental, or even prohibitive, to moving the industry forward. A few exceptions exist in New York City, where several farm operations are opening second and third rooftop locations. These farms will serve as guinea pigs, to see if organized rooftop networks (sprinkled with some progressive policy) can help propel the current niche industry into an integral facet of the urban food system.

Commercial greenhouse hydroponics,
Gotham Greens, NY

rooftop gardens [small-scale]

Whether planting a few roof-deck tomato plants or starting a community garden atop an apartment building, rooftop gardeners can't resist getting their hands dirty. These small-scale growers cultivate vegetables, herbs, flowers and sometimes even fruit to enjoy for themselves and share with others. Throughout *EAT UP*, a garden spade icon (left) will key you into relevant information for rooftop gardeners. Keep your eyes peeled!

rooftop farms [medium-scale]

Entrepreneurs, restaurateurs and urban farmers are often drawn to the commercial scale of rooftop farming. Whether growing food in soil or hydroponically, rooftop farmers must consider labor, marketing and distribution strategies in order to ensure the economic stability of their skyline farms. A fork icon (left) represents the "beans-to-bucks" approach that you can follow throughout *EAT UP*.

rooftop agriculture industry [large-scale]

The rooftop agricultural industry at large demands attention from city planners, policy makers, architects, landscape architects and academics who are interested in how rooftop agriculture can "feed the masses." A fountain pen icon (left) reminds us of the power of progressive policy and organized initiatives.

PHOTO BY ANASTASIA COLE PLAKIAS, COURTESY OF BROOKLYN GRANGE

PHOTO BY ARI BURLING, COURTESY OF VIRAJ PURI

PHOTO BY ALLEN YING

The Backdrop 11

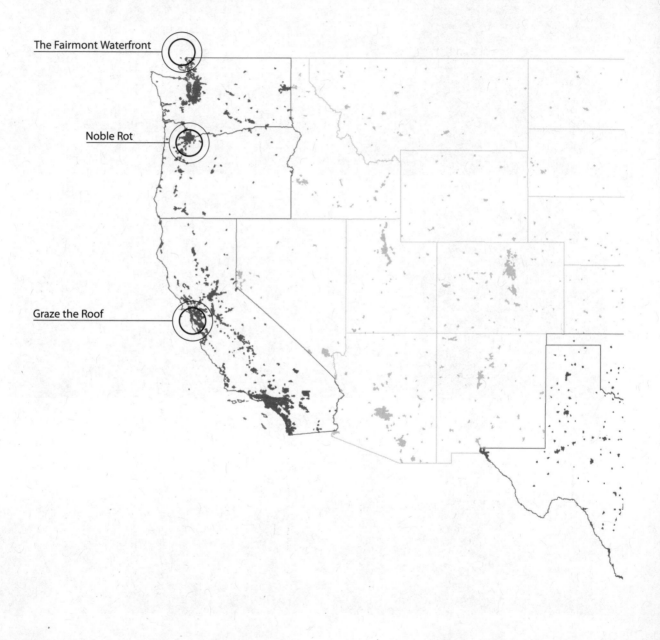

The Fairmont Waterfront

Noble Rot

Graze the Roof

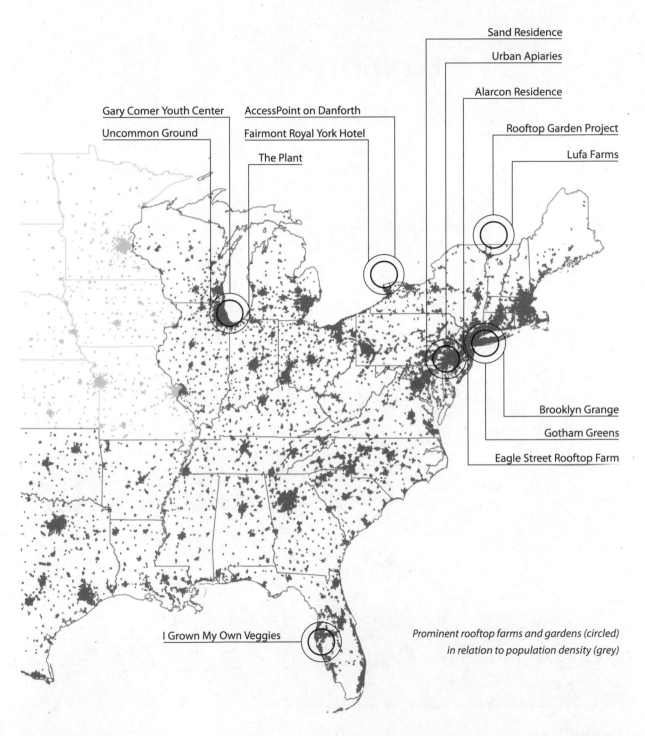

Gary Comer Youth Center

Uncommon Ground

AccessPoint on Danforth

Fairmont Royal York Hotel

The Plant

Sand Residence

Urban Apiaries

Alarcon Residence

Rooftop Garden Project

Lufa Farms

Brooklyn Grange

Gotham Greens

Eagle Street Rooftop Farm

I Grown My Own Veggies

Prominent rooftop farms and gardens (circled)
in relation to population density (grey)

The Backdrop 13

2 | What's in it for Me?

Each rooftop farm and garden, no matter how small, can enhance the lives of individuals and communities as people are drawn inward and upward. From human and social health benefits to environmental improvements, economic advantages to enhanced food access, rooftop agriculture enables and empowers people to make their communities healthier, more enjoyable places to live.

Existing rooftop (left) and artistic rendering (right),
SHARE Food Program, PA

health benefits

The many benefits associated with rooftop agriculture overlap with those of urban agriculture and green roofs. Rooftop agriculture therefore assumes the unique role of killing two birds with one stone, if you will.

human health

As with ground-level urban agriculture, food grown on roofs is as fresh as it comes. The nutrients of many fruits and vegetables degrade with time once they are harvested, which means that eating **fresh food** gives your body the nutrients it needs to take care of you. Rooftop produce is generally grown using **chemical-free** or even **organic** practices, which is great news because avoiding exposure to chemical herbicides, pesticides and fertilizers further contributes to a healthy body. Exposure to these fresh, diverse foods also encourages **healthy eating habits,** which is particularly important for children, who often shape their eating habits at a young age. Healthy eating habits (combined with regular exercise) also help combat adult and childhood **obesity.** A 2012 report from the US Department of Health and Human Services' National Center for Health Statistics found that over 35% of American adults were obese from 2009-2010.[1] The same study found that almost 17% of American children and adolescents were obese during the same period. Improved nutrition is a must.

Growing fruits and vegetables yourself offers another set of benefits. Rooftop farmers and gardeners are regularly exposed to **fresh air, sunshine, exercise,** all of which promote an overall **healthy lifestyle**. Many farmers and gardeners also report a sense of satisfaction and calm after working the soil. Rooftop agriculture creates healthy, happy people.

social health

From a social health perspective, rooftop agriculture **builds community**. It enables individuals and families to strengthen ties with each other and with their food, whether tending a community garden plot, picking up a Community Supported Agriculture (CSA) farm share or harvesting fruit and flowers at a pick-your-own farm. Community members interact with one another, experience where their food comes from, celebrate nourishment and ultimately are inspired to teach friends and family about their experience.

Furthermore, fostering relationships with rooftop farmers is essential in **strengthening the local food system**. People are much more likely to make healthy food choices and support local growers if they have access and exposure to farms and gardens. Building more urban farms and gardens, whether on the roof or the ground, provides children and adults alike with more opportunities to experience food production first-hand, make healthy, well-informed food choices and ultimately **reconnect with their food**.

Practically speaking, rooftops provide space for food production when no ground-level landscape is available. Rooftop farms and gardens are therefore particularly valuable in densely occupied cities where space is at a premium, and their construction supports the idea of building upward, rather than outward. This concept of building on unused roofs of existing structures takes full advantage of a city's **underutilized space**.

Another social health benefit of rooftop agriculture is the **provision of green space**, which is critical for building healthy communities. An article published in the Journal of Epidemiology and Human Health found that, for over 250,000 adults and children tested, perceived general health increased with the percentage of green space within the immediate living environment.[2] Another study by the same author found that 15 out of 24 physician-diagnosed disease clusters were less prevalent in test subjects living in close proximity to green space.[3] This study found the strongest correlations for anxiety disorders and depression. Additional journal articles observe positive correlations between *quality* of green space and self-reported health[4] and proximity to easily accessible green space and longevity of senior citizens.[5] In plain English, living near high-quality accessible green space (such as farms, gardens and parks) may lower anxiety and depression, improve your overall feeling of health and increase your chances of living longer. Whew!

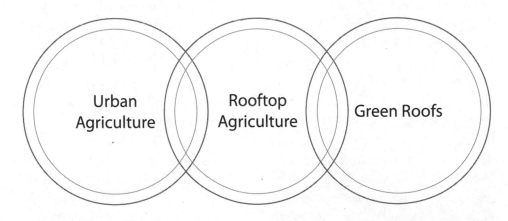

Diagram showing the overlapping benefits of rooftop agriculture with urban agriculture and green roofs

environmental benefits

Rooftop farms and gardens **mitigate stormwater runoff**, particularly when soil-intensive strategies (such as row farming) are implemented. All this soil takes the place of bare membrane roofs, which rapidly drain particulate- and pollutant-filled stormwater into the sewer system, ultimately draining into our rivers and streams. Rooftop soil and plants absorb this stormwater, dramatically slow its flow into the sewer system (thereby reducing "peak flow") and filter many contaminants from the water. Interestingly, most rooftop agricultural soils retain *more* stormwater than traditional green roofs, because agricultural soils contain more organic material, which absorbs and holds more water, than green roof media.

Urban plants are also applauded for their ability to **filter particulate**, or small grains of wind-borne sediment and dust, from the air. The leaves themselves absorb most of the particulate, which means that rooftop farms and gardens installed next to elevated highways, active factories or other point-source pollution centers may lead to a particulate-filled salad! Plants also provide the invaluable service of releasing oxygen and water vapor from their leaves. The release of water vapor is particularly important in cities, because this moisture cools the air around the plants. Water held within the soil itself is also critical in moderating microclimate, as water gains and loses heat much more slowly than bare membrane roofs and even the surrounding air. As a result, installing rooftop farms, gardens and green roofs on a city-wide scale can actually reduce the city's overall temperature. Scientists refer to the phenomenon of city-wide cooling as **heat island effect reduction**.

Rooftop row farms are also advantageous in their ability to provide insulative value to the building below, which reduces a building's utility usage and therefore the amount of fossil fuels needed to heat and cool the building. Since building insulation is a function of roof area to building volume, rooftop agriculture provides the greatest insulative benefit to large one- to two-story buildings. Industrial buildings and big box stores see the greatest insulative benefit, as evidenced by data obtained by Roofmeadow (the green roof firm where I work), from Walmart Store No. 5402 in Chicago, the world's largest green roof study site.

Lastly, reducing the distance that food travels to reach your plate **decreases fossil fuel use, air pollution and packaging**. Most rooftop farms and gardens further reduce their fossil fuel consumption by avoiding the use of chemical fertilizers, pesticides and herbicides. In the words of Ismael Hautecœur, Project Coordinator of Montreal's Rooftop Garden Project, the greatest environmental impact of all is changing people's habits.[6]

Reducing the distance that food travels to reach your plate decreases fossil fuel use, air pollution and packaging.

A chef on his Portland restaurant's rooftop,
Noble Rot, OR

economic benefits

Buying locally produced farm goods **keeps money within the local economy,** and for densely built cities that lack ground-level gardening space, rooftop agriculture may be as local as it gets! In 2003, the Maine Organic Farmers and Gardeners Association estimated that if every family in Maine spent $10 per week on local food, $104 million would enter the local economy each year.[7] This incredible statistic proves that we have the power to strengthen our local economy through strategic purchasing. Buying locally also increases the circulation speed of money, which means that dollars pass through many hands during a short period of time, thereby benefiting more people more quickly than when non-local goods are exchanged.

Rooftop agriculture also benefits the local economy by creating jobs, and at times, the need for **green jobs training.** Full-time urban farming positions may be particularly appealing to young professionals who want to work in an agrarian setting, but not at the sacrifice of an urban lifestyle. Rooftop hydroponics offers additional employment opportunities for highly trained technicians and researchers, as well as blue collar (or "green collar") workers.

Studies have estimated that, on average, produce travels 1,500 miles from field to plates in Middle America.[8] Shipping food great distances requires a middleman, who generally transfers the costs of packaging and shipping to the consumer. These costs fluctuate with fuel prices, which means that buying locally results in more stable food prices. Buying directly from farmers takes things one step further by eliminating the middleman and putting more money in the pockets of the men and women who are growing your food!

Rooftop agriculture may result in additional economic benefits to homeowners and renters, who can save money by growing their own vegetables and herbs at home. Schools, universities, restaurants, markets and even office buildings may similarly benefit by reducing their food purchase. Additionally, restaurants in particular may be able to grow specialty items, which would be prohibitively expensive to buy. The sale of food grown on location can also **bring in business,** and may enable advertising that can **increase sales.** In regions where these types of products are a novelty, the restaurant or store may be also able to charge higher prices for these highly desirable specialty items. Owners of multi-unit residential buildings may also benefit economically, by converting underutilized roof space into garden plots. In Singapore, some building owners rent these garden plots to tenants, a similar concept to charging monthly fees for a parking space, gym access or any other building amenity. In general, building owners may be able to additionally benefit from **subsidies** and **tax incentives** directed at green roofs or urban agriculture in certain cities. Owners and tenants of one- or two-story buildings may also experience **utility savings** associated with the insulative value of certain types of rooftop agriculture.

Farmer Annie Novak and her apprentices stocking an onsite farmstand, Eagle Street Rooftop Farm, NY

food access benefits

Rooftop agriculture bridges all socioeconomic gaps, thereby providing urbanites from all walks of life with **equitable access to fresh, nutritious food**. Fresh food access is particularly crucial for socioeconomically disadvantaged individuals and families, who may live in neighborhoods that lack produce markets, and who cannot afford such foods in expensive groceries outside the neighborhood. The United States Department of Agriculture (USDA) defines these **"food deserts"** as census tracts of at least 500 people that possess a poverty rate of at least 20% and contain no supermarkets or large groceries within one mile.[9] In 2012, USDA statistics revealed that, within the continental US, over 11 million Americans live in *urban* food deserts.[10]

City planners and food justice advocates routinely promote urban agriculture as a means of increasing food access within cities. When vacant lots and park land is unavailable for farming and gardening, though, rooftops and balconies may provide the only available real estate for hungry growers. In addition to space, access to information is integral to enabling rooftop gardens to develop in impoverished communities. Free or affordable Internet access (such as through public libraries) and the ability to contact knowledgeable gardeners (such as those at Cooperative Extensions) are essential pieces to the puzzle. Cooperative Extensions are offices in every US state and territory that provide free research-based information to farmers, gardeners and other parties involved with agricultural products. The extensions affiliated with Cornell University, Pennsylvania State University and North Carolina State University are particularly informative — all you need to do is call.

Food access is a **social justice issue**; it is a **human rights issue**. When urban communities lack access to affordable nutritious food, we must collectively help our fellow community members, and also educate and empower them to help themselves. Rooftop agriculture provides the canvas for food access assistance and empowerment in cities across North America. Government and non-profit food aid programs may take years to get off the ground, and may not provide fresh produce once they are fully functional. Rooftop and ground-level gardening can start today, and the bounty is delicious.

Within the continental US, over 11 million Americans live in urban food deserts.

Rooftops and balconies may provide the only available real estate for hungry growers.

A low-income neighbor volunteering in exchange for fresh produce, SHARE Food Program, PA

3 | Seed to Plate

There are many ways to grow a tomato on a roof, so what method should you choose for *your* skyline crops? This chapter describes the most effective rooftop production methods (container gardening, raised bed production, row farming and hydroponics), while considering key design criteria and the pros and cons of each strategy. The most critical take-away from this chapter is that these common ground-level agricultural practices differ significantly when applied to a rooftop. Basic principles of good gardening and farming still apply, but new considerations, such as weight, substructure drainage and wind uplift, are thrown into the mix on a rooftop.

Freshly harvested root vegetables,
SHARE Food Program's Nice Roots Farm, PA

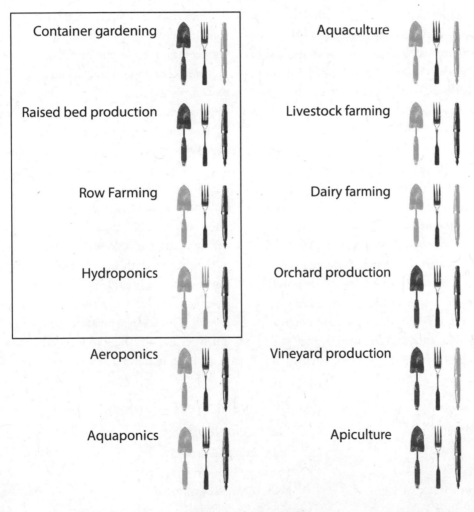

Container gardening

Aquaculture

Raised bed production

Livestock farming

Row Farming

Dairy farming

Hydroponics

Orchard production

Aeroponics

Vineyard production

Aquaponics

Apiculture

Each rooftop production method above is suitable for varying
scales of agriculture (represented by icons).
The most common rooftop methods are outlined.

1. containers

Most home gardeners focused on small-scale rooftop production choose to grow their vegetables, herbs and flowers in containers. Some plant in any vessel they can get their hands on: buckets, tins, pots, bins, hanging pouches, watering cans, kiddie pools, even old bathtubs and toilets! Once I even saw an old pair of shoes planted with flowers. Regardless of the vessels you select, container gardening is a breeze compared to raised bed production, row farming and hydroponics. It's easy enough for any novice gardener to try at home, and you may decide to start slowly with one or two containers. Rooftop container gardening is also ideal when your roof cannot sustain much weight, or you'd like to rearrange your planters frequently.

Rooftop container gardening is similar to that on the ground, except that rooftops often experience high winds. Heavy gusts can topple containers or lead to desiccation (soil drying) and winnowing (soil loss). Some rooftop gardeners position containers against a higher building wall or other windbreak to shield the plants from gusts. Anchoring the containers together can help prevent blow overs, although the soil is still susceptible to desiccation and winnowing. Lastly, container crops are more vulnerable to temperature fluctuations than those grown in contiguous soil (such as in a row farm), and the smaller the container, the more extreme this phenomenon.

Containers can be placed just about anywhere around a rooftop garden, including on tables, deck railings or a protected area of the roof deck (try using rubber walk pads or building a wooden deck). They can even hang from fence lines or overhead structures. Many container gardeners start with a few containers, and then expand their garden over time to cover as many surfaces as possible. As long as each plant receives adequate sun, water, nutrition and breathing room, you're good to go. Keep in mind that you may want

advantages ————————————

Flexible weight distribution

Mobile

Inexpensive

Manageable for novice gardeners

No digging or tilling involved

Little weeding required

Can be self-watering

disadvantages ————————————

Plant vulnerability

Soil loses moisture quickly

Soil loses heat quickly

Roots confined by container

Yields lower than other agricultural practices

Frequent fertilization required

Soil cannot be built over time

to leave some room for people in your rooftop garden! Try reserving some space for a table and chairs so that you can sit and enjoy the garden that you've worked so hard to create. Maybe leave room for a grill as well, or smaller items like bird feeders or a lounge chair. This should be a space for you, your friends and family to enjoy, not *just* a space for growing plants! Maybe spending time in your garden will even inspire others to build their own.

In general, container crops require more frequent watering than crops planted with other agricultural methods. Many

Homemade, self-watering planters ,
Sand Residence, PA

When building
or selecting
rooftop
containers,
focus on your
top priority.

gardeners choose to buy or build **self-watering containers**, which eliminate the need for daily watering. These containers are often made of molded plastic or double-stacked plastic bins and contain a false bottom that separates the soil from a water reservoir at the base of the system. A landscape or separation fabric often lines the false bottom and prevents soil from entering the reservoir, while allowing plant roots to reach the water. A watering pipe that connects the surface of the planter to the reservoir allows gardeners to periodically fill the reservoir with a garden hose. These containers conserve water by minimizing surface evaporation and allowing plants to soak up the water as needed. Gardens that contain many containers may benefit from a modular approach. **Modular containers** are those that fit together in pre-specified configurations, thereby creating a field or network of containers. Some modular containers on the market contain internal irrigation and drainage systems, which are activated when the vessels are linked together. With these systems, water typically drains out of one container into an adjacent container that is placed slightly down-slope of the first. Gauging the weights of varied pots can be challenging, but standardized containers can provide predictable loads, and therefore a means of calculating how many containers the roof can support in a given location. Homemade modular containers fabricated for row homes sometimes take

a different approach, in which a suspended frame supports various planter sleeves (see Alarcon Residence, Chapter 4).

As you have probably gathered by this point, containers come in a menagerie of sizes, shapes, configurations and materials. When building or selecting rooftop containers, focus on your top priority. Is the container's longevity most important to you, so that you can avoid replacing them every few years? Perhaps your top priority is minimizing your garden's carbon footprint through the use of recycled and reclaimed materials. Maybe a lightweight installation is paramount. What about eliminating the use of petroleum-based plastics?

In seeking a lightweight, durable, petroleum-free container, I consulted with a triple bottom line company in Upstate New York that develops strong biodegradable materials derived from fungi. Ecovative Design LLC worked with me in 2010 to develop a planter prototype made from a two-inch-thick molded material of fungal hyphae (which are similar to roots) combined with seed husks. The material was treated with a simple environmentally safe chemical, which forces the fungal material to grow a watertight, or "hydrophobic," skin that prolongs the material's lifespan. This skin enables the material to last approximately eight months (one growing season), at which point the planter can be broken down by hand and either used as a walking path material or

*Containers of all
shapes and sizes,
Graze the Roof, CA*
PHOTOS BY MICHAEL I.
MANDEL, COURTESY OF
GRAZE THE ROOF

composted to build the next year's soil. The potential for this type of material is huge. An interview with Sue Van Hook, Ecovative Design LLC chief mycologist, describes the technology in more detail.

Double-stacked plastic bins at the Sand Residence, PA
Photo by Michael I. Mandel, courtesy of Graze the Roof

materials + sizes

Planting containers are commonly made of plastic, metal, wood, ceramic or fiber. The materials vary in strength, durability, longevity and color, as well as recycled and recyclable content. Regardless of material, light-colored containers absorb less heat than dark containers, which causes the soil to dry out less quickly. Similarly, large containers retain more soil moisture than small, and plastic containers retain more than unglazed ceramic pots. Plants in large, light-colored plastic containers therefore require the least frequent watering. Each crop requires a different soil depth to sustain its root mass. Use 1- to 3-gallon containers for herbs, salad greens, radishes, green onions, chard and basil, and 4- to 5-gallon containers for deep-rooted crops like tomatoes, cucumbers, eggplant, beans, peas, cabbage and broccoli.

garden soil

Your average off-the-shelf potting mix contains a combination of organic matter (such as peat moss, sawdust or bark) and mineral material (such as vermiculite, perlite, pumice, sand or calcified clay). This type of soil is perfectly suitable for container gardening, so long as it remains porous enough to enable proper drainage to occur. In properly draining systems, the soil's organic matter is what retains most of the moisture, and consequently its weight. On rooftops where lightweight growing media is essential, experiment with an increased mineral content.

More advanced gardeners may choose to mix their own soil recipe from scratch, which can be both fun and challenging!

nutrients + drainage

Your container plants will be thirsty, and hungry, too! Watering requirements vary with container size, container configuration, microclimate and climate. Some individual crops even need more water than others. For vegetable crops, the soil should remain moist, but not overly saturated. Plant roots can suffocate and die in the absence of proper drainage; holes through each container's base enable drainage to occur. A gravel reservoir at the base can help promote drainage as well. As a rule of thumb, water container crops twice per day in hot, dry climates, once per day in temperate and cold climates, and one or two times per week when using self-watering containers. All this watering will flush nutrients out of the soil, causing a need for more frequent fertilization than with raised beds or a ground-level garden. Try feeding your plants with compost, the juice from compost (called "compost tea") or organic fertilizer, and avoid chemical fertilizers whenever possible.

biodegradable container

innovative approach

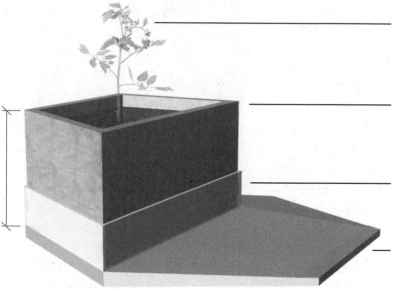

planting
species and spacing appropriate
to soil depth and shade conditions
of planter

planter sleeve
molded hydrophobic fungal hyphae-
seed husk insert, begins biodegrading
into growing media after 8 months
with physical disturbance

sleeve support
ultraviotet (UV) stabilized recycled
HDPE plastic with weep holes for
drainage

sloped paver
light-colored molded recycled tire
rubber with adjustable HDPE plasic
pedestals, runoff from drainage and
rain at unit's depressed edge

18″

12″

6″

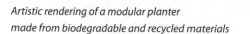

*Artistic rendering of a modular planter
made from biodegradable and recycled materials*

Sue Van Hook interview
Ecovative Design LLC, Chief Mycologist

LM Can you describe the flower pots your company developed?

SVH The pots are made out of a seed husk-fungal hyphae material that is completely compostable and biodegradable .... We tried [coating the pots with] a lot of different waxes — beeswax, paraffin, etc., but the pots quickly biodegrade.

LM Is there a longer-lasting material that can be used to coat the pots?

SVH Yes, we've experimented with a chemical treatment that changes the chitin back to chitosan to create a hydrophobic skin so that the container can hold water .... **It's water-tight for the short term,** but we haven't tested how long it can hold water for in the long term .... My guess is it will last for eight months.

LM What is the potential for a field application of the chemical coating?

SVH You could reapply the chemical in the field, but you could also replace the pot after the growing season and **reuse** the existing one as **mulch or a walking surface.**

LM Are there structural or size limitations to the material?

SVH The largest tested size has been four-foot-by-eight-foot panels used for wall insulation at a school in Vermont .... [The material is] extremely hard.

LM Do you think that the hyphae could come from fungi grown on the roof?

SVH Oh, yes. One of the modules could be a block of spawn that fruits .... [The module] must be shaded and moist. It could be located in the shade of an adjacent building.

LM How would you grow this type of fungi on a roof?

SVH Oyster [mushrooms] are grown on a straw block. It takes two weeks to grow the block up, and then you could insert the block. [But] oysters are very short-lived. Shitakes last longer, and for growth, they require a log or a log fabricated out of sawdust .... You could inoculate the log in the spring and collect the fruit in the fall.

LM Can adhesives be used on the material?

SVH Yes, any adhesive can be used.

"seed husk-fungal hyphae material that is completely compostable and biodegradable"

"One of the modules could be a block of spawn that fruits."

2. raised beds

If higher yields are what you're after, raised beds might do the trick. These are built-in-place or prefabricated low-profile structures filled with soil, which provide more contiguous growing space than containers. Raised bed production is ideal for rooftop gardens and farms for which a relatively simple and inexpensive installation is desired. Relative to other growing methods, these growing areas are: a) more permanent than a container garden, and b) more lightweight than a row farm. Raised beds are also ideal for rooftop community gardens, and for growing almost any crop — including root vegetables like beets and carrots, and invasive perennials like strawberries.

Rooftop raised beds are similar to those on the ground, except that they must be designed to accommodate a building's rooftop weight restrictions. This means that thin beds (often less than 12 inches deep) made of lightweight materials (such as wood or metal siding) are ideal — brick and cinder block are rarely used. The beds may need to be located farther apart than ground-level raised beds in order to distribute their weight, or placement along the parapet or over columns (where a roof is generally strongest) may be required. In order to prevent serious structural damage or a full-blown roof collapse, a licensed structural engineer must evaluate

advantages

Simple + attractive

Better soil drainage than containers + row farming

Higher yields than containers
    (when planted intensively)

Decreased soil compaction + erosion

Soil gains heat more quickly than row farming

Extended growing season (during spring)

Easy weed control (due to intensive
    planting shade)

Ideal for root crops

Contains aggressive plants

disadvantages

Limited production area

Susceptible to temperature fluctuations

Soil dries out more quickly than row planting

Occasional frame maintenance

Cost of construction and repairs

Frequency of replacement
    (wood-framed beds)

Weight (relative to containers)

Elevated cedar raised beds with steel framing,
Uncommon Ground, IL

each roof and specify the locations in which raised beds can safely be located.

Unlike their ground-level counterparts, rooftop raised beds are designed to drain more efficiently, so as not to pool water and exceed weight limits. Drainage can be achieved by installing each bottomless raised bed frame over a strong perforated layer such as a synthetic sheet drain with a high compressive strength (a common green roof material) or perforated plastic pallets. It is important that the material that you select for this layer is relatively durable, and will last at least as long as the raised bed frame itself. Four- to six-ounce landscape fabric, or separation fabric as it's called in the green roof industry, should be placed over the sheet drain and up the frame's walls, to prevent soil from clogging the drainage layer. Water will drain vertically through the soil, and then horizontally through the sheet drain layer to the roof deck. The whole assembly should be installed on top of a durable material (like 30 mm polyethylene sheeting) to protect the roof's waterproofing membrane from abrasions.

Accessing a roof to replace deteriorated materials can be difficult and, at times, prohibitively expensive (particularly if a crane is needed). Rooftop raised beds therefore generally utilize more durable materials than those on the ground, despite higher costs.

Raised beds can be built out of various materials, but on rooftops, most seem to fall into the categories of either wooden or steel-framed. **Wooden raised beds,** such as those atop the Portland, OR restaurant Noble Rot (opposite page), are constructed of non-pressure treated wood with hardware at the corners. When using softwood (such as pine), this type of raised bed is inexpensive to build but needs to be replaced after several growing seasons as the wood deteriorates. Beds made of hardwood (such as cedar) will last longer, but can be extremely expensive. **Steel-framed raised beds,** such as those on top of Chicago's Uncommon Ground restaurant (see Chapter 5), are more durable than softwood beds, but also more expensive to install. The welded steel framing not only looks smart and sophisticated, but functionally it prevents the wood from warping. The framing also allows the beds to be elevated, which is beneficial for the farmer's back and helps to hide the drainage and irrigation equipment underneath the beds. Thin or lightweight elevated raised beds are susceptible to tipping over in heavy winds. Consequently, some designers secure them to the decking to avoid "wind uplift." Heavier raised beds are less susceptible to wind uplift, even in the absence of anchoring.

In addition to growing space, many rooftop raised bed gardens and farms contain gathering areas that accommodate touring visitors and community dinner events. These spaces are particularly useful for rooftop farms over restaurants.

In addition to growing space, many rooftop gardens and farms contain gathering areas.

Additionally, outdoor classrooms, demonstration plots and seating can enhance rooftop gardens that are oriented toward educational and community outreach.

Wooden raised beds with rubber walk pads,
Noble Rot, OR
PHOTO BY JOHN Q. PORTER, COURTESY OF NOBLE ROT

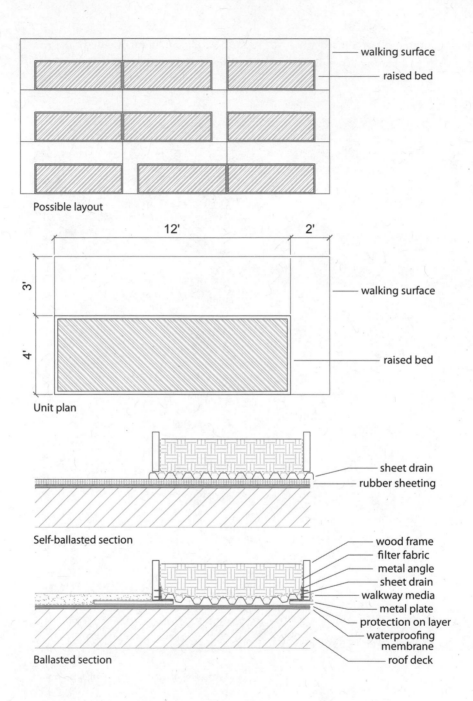

walking surface

raised bed

Possible layout

12'   2'

3'

4'

walking surface

raised bed

Unit plan

sheet drain
rubber sheeting

Self-ballasted section

wood frame
filter fabric
metal angle
sheet drain
walkway media
metal plate
protection on layer
waterproofing
membrane
roof deck

Ballasted section

dimensions

Most raised beds are four feet wide when accessible from both sides, and three feet wide when accessed from a single side. Twelve- and 24-foot beds are common, but length is ultimately determined by site layout and material dimensions. Paths between the beds are commonly three feet wide, just enough to fit a wheelbarrow through.

Raised bed height can vary within a single rooftop garden or farm, particularly when structural limitations do not allow deep soil in certain areas. Shallow beds that are four inches deep can accommodate most salad greens and herbs. Six-inch beds should accommodate the rooting zones of most Brassicas (e.g., broccoli, cauliflower, kale) and leafy greens like collards. Twelve-inch beds can sometimes sustain Solanums (e.g., tomatoes, eggplant), when they are planted less densely than usual. Ideally, 18-inch beds are used for Solanums, as well as root crops like beets and even carrots.

materials

Rooftop raised beds can be built with new or salvaged materials — the more durable, the better. Some materials that are used for ground-level raised beds (concrete blocks, brick, stone) are less suitable for rooftops, given their weight. Pressurized wood and railroad ties contain dangerous chemicals that can leach into certain food crops, and should always be avoided. What materials are left? Untreated wood and metal paneling. The former is lightweight, easy to work with and available in standard sizes. The drawback is that untreated wood deteriorates quickly, and generally needs to be replaced after two to four years, so plan for replacement accordingly.

Most raised bed growers mix their own soil, which is generally some combination of compost, peat moss and vermiculite or perlite. The synthetic sheet drain underneath the framing can be protected from hoes and shovels by installing a hard plastic grid, known in the green roof industry as a "shovel guard."

inputs

As with containers, your raised bed crops will need water and nutrients. Hand watering is often ideal for a one- or two-bed operation, while surface drip irrigation will save you time in larger gardens. Regardless of what type of irrigation you opt for, surface mulching (with straw or even shredded newspaper) will generally help to reduce the rate of desiccation. Just be sure to pick a material that doesn't blow off the roof! In general, rooftop raised beds are much more susceptible to desiccation than those on the ground, due to the wind exposure on roofs.

Regular applications of organic fertilizer are often necessary to maintain plant health in raised beds. Adding compost is also beneficial for restoring nutrients and building soil health, so long as the compost does not cause the raised beds to exceed the weight limit set forth by your structural engineer.

Charlie Miller interview
Roofmeadow, President

LM Where did you first learn about green roofs?

CM I learned about them from a colleague of mine from Germany named Joachim Tourbier, who was a professor at the University of Pennsylvania. He arranged for me to have interviews with the two primary [green roof] companies in Germany.

LM Did the green roof industry exist in the US when you came back from Germany?

CM There was no green roof industry [when I got back in 1997] …. Of course, prior to that there had been the whole earth shelter movement in the 1960s.

LM Have you ever worked on a farm?

"I've been involved with dirt and water and plants ... for a long time."

CM Yes. I've been involved with dirt and water and plants in one form or another for a long time.

LM Do you have a formal education in soil science?

CM I have a master's degree in geology and geophysics. I also have a master's degree in civil engineering with an emphasis in physics and soil mechanics …. [After careers in farming and geology] I went into civil engineering, and was involved in urban storm-water management, small dam investigations, water supply development, contaminated groundwater remediation and landfill design.

LM To what degree do you direct the engineering of rooftop soils?

CM [At my green roof firm, Roofmeadow] we are continually in the process of evaluating how different types of media have performed in different parts of the country. We continually evaluate the soil characteristics to maximize performance and reduce cost. **The proper selection of media is the most important decision to make.**

LM Have you ever written a specification for rooftop agricultural soil?

CM I have, but I don't consider myself to be an expert. My opinion is that agriculture is a very diverse activity, and the **types of media are going to be very varied depending on the project.** The one-size-fits-all perspective for rooftop agriculture is even less likely that the one size for green roofs …. Religion and agriculture are both fields where subjective experience trumps science. There are too many variables.

LM How did the rooftop agricultural soil differ from traditional green roof media?

CM It had higher organic content. Most green roof soils contain 4% to 8% organic matter …. This agricultural soil contained **20% organic matter,** which is really high. We also allowed for lower porosity …. The target was really creating a something that fell between green roof media and nursery soil.

LM What material do you prefer to use to frame rooftop raised beds?

CM I would go with Trex or an environmentally friendly, semi-durable material …. This **isn't agriculture in the normal sense** …. You make material choices that are more based on aesthetics and environmental messaging.

LM What is the best way to protect rooftop raised bed crops from wind damage?

CM You want **low-stature crops** …. You would want to plant them more densely than ground crops so that they can protect each other from the wind.

LM What potential is there for rooftop agricultures to positively impact communities?

CM It's great at acting as a vehicle for getting communities to cooperate, for people to learn community organizing skills and to **familiarize communities with the look and the feel and the taste and the attractiveness of fresh produce.**

LM What are the most significant barriers to the success of a widespread rooftop agricultural movement in the US?

CM The same as there is to agriculture everywhere in the United States. The cost of food in the US is the only commodity that has steadily decreased in absolute price for the last 100 years. The price we pay for food today is based on economies of scale and cheap labor. You don't want an urban agricultural movement to propagate those evils into a city. You want to think that rooftop agriculture can provide a living wage and work on a human scale that provides an intimate relationship between people and their food. It seems to me that the economic barriers to this are insurmountable.

LM Would you grow food on your own roof?

CM Yeah. I absolutely would because I can't grow anything on the ground there.

> "Material choices that are more based on aesthetics and environmental messaging."

3. row farming

Row farming, or linear crop production in contiguous beds, is the primary agricultural method practiced by American farmers, and is suitable for almost every climate found within the continental US. On rooftops, row farming is ideal for large-scale production, when moderate to high yields and high stormwater management benefits are sought. Row farms allow for a flexible bed layout that can be rearranged when necessary. This type of production can support most any type of crop, although low-yielding crops (like corn and grains) and large, spreading crops (like pumpkins) should generally be avoided.

Row farming on rooftops is fairly similar to traditional ground-level production. The contiguous soil system in both scenarios allows for uninterrupted hydrologic and microbial activity. Most commercial rooftop row farms in the US tend to practice organic polyculture farming, whereby many crop varieties are grown throughout the season without the use of chemical fertilizers, herbicides or pesticides. Just like on the ground, many of these rooftop farms develop composting programs so that soil can be built over time. Some farms additionally integrate chickens (for pest management), bees (for pollination) and rabbits (which produce organic manure).

Rooftop row farming requires buildings with a structural capacity strong enough to support extreme loads. The soil may be deep, and the high level of organic matter, which is common in these farms, will retain water and add to the load. A 12-inch-deep area, for example, will weigh 84 pounds per square foot at the very least. This means that the most suitable structures are likely existing concrete deck industrial buildings or new structures that are built with heavy loads in mind. Rooftop row farms can be constructed in several different ways, but the system will always be built over top of a building's

advantages

Unobstructed flow of water, roots and microbes

Stable soil moisture + temperature

Opportunity to build soil each year

Flexible layout

Ability to plant almost any crop

Potential for large-scale production

Increased efficiency from scale

disadvantages

Weight

Soil may require significant amendments

More difficult to plant intensively

Possible weed problems

Possible mixing of path material + media

Walking path compaction

waterproofing membrane. While some practitioners may prefer one waterproofing material over another, the truth is that a rooftop farm, like a green roof, can be built on top of almost any type of membrane. Depending on what membrane is used, a plastic (usually polyethylene) root barrier may be required to prevent the plant roots from damaging the waterproofing. Either a coarse aggregate or synthetic drainage sheet will be installed to ensure proper drainage, and a filter fabric will separate the drainage layer from the soil.

*Mounded crop rows in full production,
Eagle Street Rooftop Farm, NY*

layout

Solar orientation, wind screening, water access, degree of mechanization and desired crops are essential to consider when designing a rooftop row farm. Soil depth is largely determined by structural and cost limitations. The length of the rows themselves can vary depending on site dimensions, whereas row width is generally more standardized. Farm rows are typically three to four feet wide for most crops, with one-to-two foot wide paths between rows. Wider paths can be installed occasionally, although rooftop acreage is expensive, and should be used wisely. Be sure to install edging to separate the walking path material from the beds themselves.

In addition to crop rows, rooftop row farms often reserve space for compost piles, beehives, tool and storage sheds and sometimes an observation platform. Additional space off-site is often valuable for washing and processing produce, and possibly setting up a market.

planting strategies

Strategies such as crop diversification, intercropping, cover cropping, crop rotation, succession planting and relay planting can all lead to greater yields in organic farms. Succession planting in particular can be beneficial in ensuring a continuous harvest. It is generally most successful when determined before the growing season begins for the year.

Since rooftop space is at a premium, some farmers choose to focus on a narrower crop selection: those that are highly profitable, or those that can be used as key ingredients in value-added products, such as hot sauce or salsa. Hot peppers are becoming a popular rooftop crop specifically for this reason. Care must be taken, of course, to vary the planting year to year, so as to maintain soil health and structure.

inputs

Irrigation is a must for row farming, particularly when larger production areas are at play. Since crops require one inch of water per week on average, irrigation drip lines are often preferable to hand watering. Surface drip lines can be installed inexpensively, and when positioned at the base of each plant, will conserve water as compared to hand watering. Burying the drip lines two inches below the surface of the media conserves even more water (and encourages deeper root growth), but buried lines may make hoeing and planting difficult. Regardless of the type of irrigation that is deployed, mulching with straw or other materials can reduce the rate of soil moisture loss.

Organic fertilizers, pesticides and herbicides are necessary on most rooftop row farms, as is integrated pest management (IPM). This practice involves attracting beneficial insects to prey on pest insects that attack the crops.

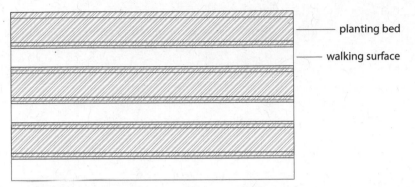

planting bed

walking surface

Layout Plan

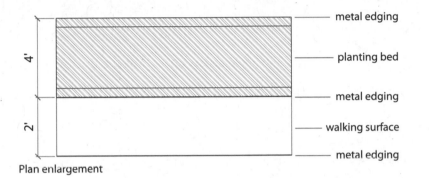

metal edging

planting bed

metal edging

walking surface

metal edging

4'

2'

Plan enlargement

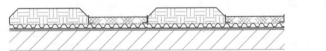

Section

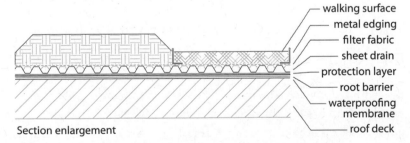

walking surface
metal edging
filter fabric
sheet drain
protection layer
root barrier
waterproofing
membrane
roof deck

Section enlargement

Lisa Goode interview

Goode Green, Principal + Eagle Street Rooftop Farm Landscape Architect

LM Was the building [below Eagle Street Rooftop Farm] structurally reinforced prior to construction?

LG No.

LM What type of waterproofing membrane was specified?

LG We installed atop the existing roof which was an old-style tar roof.

LM Was a drainage layer installed?

LG Yes, we used an Optigrun base system — polyethylene, fabrics, drainage mats, etc.

LM Why was row farming implemented rather than raised beds?

LG Because **the farm is a green roof** [with a continuous soil and drainage system].

> "The biggest design challenge was installing to the engineer's limit — 40 psf."

LM What was your biggest design challenge?

LG Installing to the engineer's limit — 40 [pounds per square foot]. It would have been nice to [install] more soil.

LM What was your biggest construction challenge?

LG None — this was an **easy installation**

LM What was the rooftop farm's total construction cost?

LG $60,000 for the green roof [and] about $3,000 for the farming materials and seeds.

LM How long is the expected payback period?

LG With the amount of volunteer hours [which amounted to] 85% of the farming, it would be hard to gauge an expense and then a payback. The farm is not a viable income for the main farmer and her few partners. **[Eagle Street Rooftop Farm] is a center for education and advocacy** supporting urban farming, green roofing and fresh produce in under-served neighborhoods.

LM What was the most significant construction expense?

LG Edging [was the most expensive constriction expense], but for this project we got free rafters which we recycled for use on the roof.

LM Were alternate methods (besides a crane) considered for transporting materials to the roof?

LG No. The location was incredibly easy for a **crane installation.**

LM How is the roof accessed for daily activities?

LG Stairs.

LM How much demand have you encountered for rooftop agriculture in NYC?

LG Lots — due to all of the press from last summer [2009], we are getting a lot of calls.

LM Is rooftop farming in NYC appealing to clients from multiple socioeconomic groups?

LG Yes.

LM Do you think that large-scale (widespread) rooftop agriculture is viable in NYC?

LG Yes, but **not for any type of profit.** The scale is just not possible [without considering replicability across many roofs to achieve greater acreage].

LM Do you think that rooftop farming is a fad that may pass?

LG Possibly, although not without another type of system that delivers fresh produce.

"Edging [was the most expensive construction expense]."

"The location was incredibly easy for a crane installation."

4. hydroponics

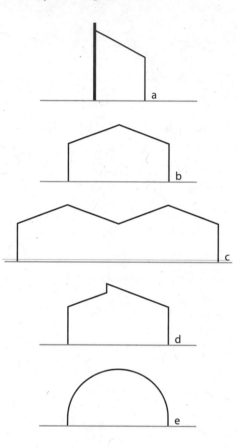

If you want to grow as much food as possible in a limited space, consider hydroponic production, a method of growing plants in a nutrient solution, without soil. Most hydroponic systems provide plants with an inert substrate in which to root (instead of soil), such as coconut coir, mineral wool, perlite or expanded clay. Using this production method, Gotham Greens, a hydroponic company in New York City (see Chapter 6), produces 20 to 30 times as much produce per acre as a typical ground-level row farm.[1] Lufa Farms, a hydroponic farming company in Montreal, QC (see Chapter 5), similarly grows 10 to 15 times the produce as a soil-based farm.[2] These yields may sound like magic, but producing them requires sophisticated and a high degree of skill.

In cold to moderate climates, hydroponic facilities are generally housed within **greenhouses** that carefully monitor

advantages

High yields

Lightweight (compared to soil)

Efficient use of space

Controlled growing conditions (greenhouse)

Extended growing season (greenhouse)

Off-season harvesting of cold-season crops (greenhouse)

Creates high-skilled jobs

disadvantages

High initial investment

Lengthy construction period

Regular maintenance required

Energy intensive

Difficult to manage stormwater (relative to open soil system)

Requires high-skilled labor

Highly calibrated hydroponic greenhouse,
Gotham Greens, NY

and calibrate temperature, humidity, air circulation, nutrient output and even light. Lufa Farms developed an impressive greenhouse that contains multiple microclimates that cater to the needs of each crop. This and other high-tech greenhouse facilities typically recirculate water through the hydroponic system, and sometimes incorporate harvested rainwater from the greenhouse roof to further conserve water. The recirculating irrigation system at Gotham Greens, in fact, uses 20 times less water than conventional agriculture, according to the company's website.[3] By calibrating indoor climate, greenhouses extend the growing season for many crops. In certain climates (including Montreal), year-round production is even possible with the use of highly insulative glass and innovative heating measures.

Greenhouses occur in a variety of styles: lean-to (a) greenhouses consist of south-facing glass placed against an existing wall; even-span (b) structures possess two roof slopes of equal pitch; ridge and furrow (c) greenhouses contain multiple A-frame spans that are connected along the eaves; sawtooth venting (d) structures possess a vent along the eaves where one pane overlaps another; and Quonset (e) structures, which are specific to plastic hoop houses.[4] Commercial rooftop greenhouses generally exhibit even-span or ridge-and-furrow construction. As compared to ground-level greenhouses, those on rooftops are highly engineered to withstand extreme winds and snow loads. The structure's steel framing is generally bolted to the roof deck and sealed around the base to prevent air from forcing its way underneath the greenhouse and causing wind uplift and shaking.

In warmer climates (such as southern Florida, parts of Asia and the tropics), year-round hydroponics is often possible without the protection of a greenhouse. This practice, known as "**open-air hydroponics**," generally involves stacked lightweight containers filled with an inert substrate. Irrigation often occurs from a feeder pipe that connects the stacks from the top and supports drip lines that water the uppermost container in each stack. The containers have holes at their base, allowing excess water to leak into the container below. These systems often recirculate the nutrient solution to conserve water. In contrast to ground-level operations, rooftop open-air hydroponic farms are often designed to protect plants from high winds and extreme sun. I Grow My Own Veggies, a 3,000-square-foot open-air hydroponic farm in Sarasota, FL (see Chapter 6), addressed these issues by installing tennis court screening. Even still, the sun is too strong for certain crops during the summer.

Hydroponic farms designed for maximum yields leave little room for non-production programming. There may be room for a main walkway with interpretive signage, where school groups can observe hydroponic farming firsthand.

Open-air hydroponic tomato plants,
I Grow My Own Veggies, FL

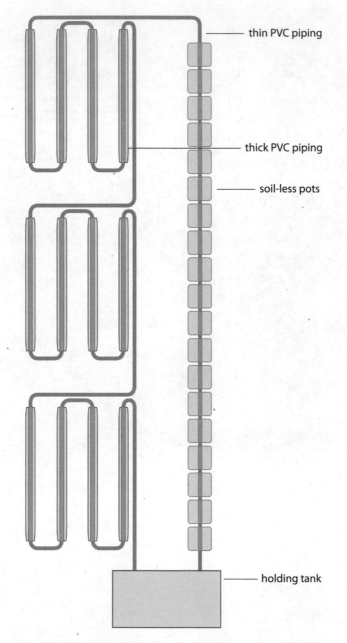

thin PVC piping

thick PVC piping

soil-less pots

holding tank

Greenhouse hydroponic layout diagram

solar orientation

Solar orientation is critical in positioning a greenhouse so that plants receive as much sunlight as possible. Direct southern exposure is optimal, with effectiveness decreasing as the building angle shifts away from an east-west orientation. When selecting a building on which to site a rooftop greenhouse, pick one that exists on a true north-south grid. Also look for a roof with no taller buildings or vacant lots to the south. Taller buildings and future buildings can shade your greenhouse.

Solar collectors can help increase the air temperature within a passive solar greenhouse. These heat sinks are located in direct sun within the greenhouse, and are often built into an insulated north-facing wall. A solid brick wall, in fact, makes an excellent solar collector, as does a wall of stacked water-filled plastic jugs.

energy use

Hydroponic rooftop greenhouses typically require fewer energy inputs than their ground-level counterparts. When growing crops year-round, much of a greenhouse's energy use comes from heating during the winter. On a roof, greenhouses may take advantage of heat escaping from the building below. A production greenhouse on top of Eli Zabar's Vinegar Factory in Manhattan, for example, is warmed by heat ducts from the bakery below that pass through the greenhouse. Additionally, most rooftop greenhouses in North America are relatively new and highly engineered to use as little energy as possible. Photovoltaic (PV) technology further allows hydroponic rooftop farms to produce their own renewable energy right on the roof, next to the hydroponic facility. Ground-level farms do not always have this luxury. Siting a greenhouse near consumers further reduces a farm's energy consumption, by minimizing transportation and potentially packaging as well.

pest control + pollination

Hydroponic rooftop facilities sometimes incorporate Integrated Pest Management (IPM) into their farm practice to battle harmful insects without the use of chemicals. This trend seems to be more common on rooftops than in ground-level facilities, perhaps because the messaging of most rooftop hydroponic farms is that of chemical-free production methods.

Most rooftop greenhouse farmers also introduce beehives into their facilities, in order to ensure that their crops are pollinated. Using IPM in place of pesticides helps to protect the resident pollinators from becoming poisoned.

integrated approach

production method comparison
to help you plan

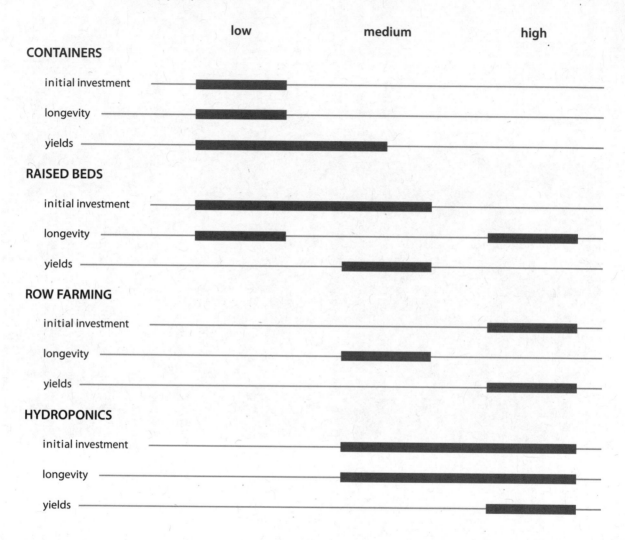

	low	medium	high

CONTAINERS
- initial investment
- longevity
- yields

RAISED BEDS
- initial investment
- longevity
- yields

ROW FARMING
- initial investment
- longevity
- yields

HYDROPONICS
- initial investment
- longevity
- yields

With so many agricultural strategies to choose from, why pick just one? Many rooftop farmers and gardeners deploy several production methods in order to grow the crops they want within the parameters of the given space, budget and structural capacity.

This diagram (left) compares rooftop containers, raised beds, row farming and hydroponics in terms of initial investment, longevity of the infrastructure and typical yields. It is meant to capture general trends, but of course outliers do occur. As you can see, **container gardening** is generally the least expensive production method, but its infrastructure is more temporary and yields are generally low. Planting in self-watering containers is arguably the most effective way to increase yields within the bounds of a container. **Raised beds** generally require a slightly higher initial investment than containers, but yields are also higher. The longevity of the beds themselves is variable, given that untreated pine may last only two years, while metal-framed cedar beds may last upwards of ten. Rooftop **row farms** require significantly more upfront capital than containers and raised beds, but in turn may produce higher yields. The underlying layers within a row farming system are durable and long-lasting, but the planting beds themselves require continuous upkeep. The media, too, requires regular inputs to remain fertile enough for vegetable production. For extended longevity, try not to fall behind with upkeep. **Hydroponics** occupies the opposite end of the spectrum from container gardening. This production method is expensive, its infrastructure long-lasting and its yields through the roof: high input, high output.

Now that you know how these production methods compare, think about which approach would be best for your project. Is there any latitude in installing more than one method? For school, religious institution, community center and apartment building roofs, try using both containers and raised beds. For universities and restaurants, maybe a row farm with a few raised beds will do the trick.

Regardless of which methods you select, be sure to research local building and zoning codes, and always consult with a licensed structural engineer. Chapters 4 through 6 lay out these steps in detail, so read up and get started!

crop planting guide

to whet your appetite

	arugula	basil	beets	bok choi	broccoli	brussel sprouts	cabbage	carrots	cauliflower	celery	chard	chives	cilantro	collards	cucumbers	daikon	eggplant	endive	fennel	garlic	green beans	kale	kohlrabi	lavender
CONTAINERS																								
1-3 gallon	X	X									X	X	X					X		X				
3-5 gallon				X	X	X	X		X					X	X	X	X		X		X	X	X	X
RAISED BEDS																								
4" deep	X											X	X					X						
12" deep		X		X			X				X			X	X		X				X	X	X	
18" deep			X		X	X		X	X	X					X	X	X		X			X		X
ROW FARMING																								
12" deep	X	X		X							X	X	X	X	X		X	X			X	X	X	X
18" deep			X					X	X					X	X	X			X			X		
HYDROPONICS																								
Open-air	X	X		X			X				X	X	X	X		X	X	X				X		
Greenhouse	X	X		X							X	X	X	X	X	X	X	X				X		

leeks	lettuce	mint	mizuna	mustard greens	napa cabbage	okra	onions	parsley	peas	peppers	pole beans	potatoes	radicchio	radishes	rosemary	runner beans	rutabaga	shallots	snap peas	snow peas	sorrel	spinach	squash	sweet potatoes	thyme	tomatoes	toma tillos	turnips	zucchini
	X	X	X	X				X					X	X				X			X	X			X				
X					X	X	X		X	X	X			X	X	X		X	X			X				X	X		X
	X		X					X				X	X								X	X		X					
X		X		X	X				X	X					X		X	X	X	X						X	X	X	X
						X	X			X	X	X					X						X	X		X	X		X
X	X	X	X	X			X	X	X			X	X	X	X		X	X	X	X	X	X	X	X	X	X	X	X	X
						X	X		X	X	X						X						X	X		X	X		X
X	X		X	X			X	X		X		X				X					X	X		X	X	X	X		
	X	X		X	X				X			X	X	X							X	X				X	X		X

Certified organic rooftop radishes,
Uncommon Ground, IL

the argument for organics

With so many tasty crops to choose from, consider going organic! The US Department of Agriculture (USDA) defines organic food and agricultural products as those that are produced in the absence of genetic engineering, irradiation, sewage sludge, synthetic fertilizers, pesticides and herbicides.[5] Organic production must "foster cycling of resources, promote ecological balance and conserve biodiversity," and cannot be granted until three years after the last synthetic chemical or sewage sludge application. Organic certification in the US can only be granted by a USDA-accredited "certifying agent," who is granted permission to inspect the site. Farmers seeking organic certification must maintain detailed records of their operation, submit an updated organic production plan annually and can only advertise and label products as organic after gaining USDA approval.

In October 2008, Uncommon Ground proudly became the nation's first certified organic rooftop farm. This 2,500-square-foot (0.06 acre) rooftop gem provides roof-fresh produce to the sustainably minded restaurant below the farm (see Chapter 5). Many other small farms and gardens cannot afford the fees associated with organic certification, but commit to raising crops with organic methods anyway, for good reason! Large rural farms frequently grow acres of a single crop (such as corn, soybeans or wheat), which require chemical inputs to keep diseases and pests at bay. Planting diverse crops within a given space, a practice called polyculture, reduces the chances of disease and pests, thereby minimizing the need for harmful chemicals and genetically modified crop varieties. Polyculture is commonly deployed in organic farms and gardens, as are other practices such as IPM, crop rotation and companion planting. By default, most urban gardeners and farmers practice polyculture in order to grow the crops that "eaters" demand. After all, why would you grow only eggplant in your garden? How boring! Urban growers often choose to avoid chemical inputs for a variety of reasons. Most commonly because they are in direct contact with their food and do not wish to apply harmful chemicals that they will ultimately end up eating.

Some hydroponic rooftop greenhouses, such as Lufa Farms in Montreal (see Chapter 5), practice organic growing methods but are not certified organic. Unfortunately, the North American organic certifying entities do not yet recognize hydroponics as an organic method of production, even when IPM is used and chemical inputs are avoided.

If you're on the fence about whether or not to "go organic," then try talking with some urban growers in your area about their experiences. They may even reveal some helpful tips that can help you succeed within your city.

milk, eggs + wool

Integrating animals into a vegetable farm is nothing new. For centuries, farmers have valued poultry (like chickens and turkeys) and livestock (like sheep, goats and cattle) for their ability to **maintain soil health.** These animals provide the invaluable services of removing pests from the soil, keeping grasses and weeds at bay and returning nutrients to the soil (through manure deposits), thereby reducing the need for fertilizers, pesticides and even mowing.

Other products from rooftop animals include **eggs, milk, meat, poultry,** and **wool.** The industrial-sized egg-laying operations common in rural settings are not suitable for rooftops. Most rooftop farms deploy a handful of free-range hens to patrol the vegetable beds and deposit manure. At some farms, they are kept in a rooftop coop, which may be mobile. As with many urban farms that keep chickens, a diversity of breeds is typically present on rooftops. When contemplating the addition of chickens to your rooftop farm, check with local zoning and municipal codes to see if the animals are permitted within city limits. Roosters are almost always banned, so focus your energy on the ladies!

In addition to chickens, one New York-based rooftop farm houses rabbits. These cuddly hoppers are raised for their manure, which adds valuable nutrients to the soil. In contrast to horse, cow and chicken manure, which needs to be aged or composted before use around crops, rabbit droppings can be applied directly from rabbit hutch to planting beds as a "cold manure." When introducing rabbits to a rooftop farm, be sure to protect your crops from the fuzzy nibblers.

On rooftops where vegetables, chickens and rabbits are not of interest, goats may serve as ideal occupants. The grassy roof of Al Johnson's Swedish Restaurant and Butik, in Wisconsin, is home to a herd of grazing goats that attracts a national audience. In a 2007 interview, restaurant owner Al Johnson explained that sod roofs are common in Sweden (where he was born) as a means of moderating indoor building temperatures.[6] In 1973, Johnson decided to install a sod roof atop his log cabin restaurant, which was built with six inches of soil over a plastic drainage layer (known in the green roof industry as a reservoir sheet). Rather than mowing the roof, Johnson introduced a herd of goats to keep the lawn at bay. By training the pack leader to stay on the roof, the others follow suit and generally refrain from jumping. The iconic restaurant draws large crowds during the summer months, and according to Johnson's son Rolf, the goats are responsible for putting the village of Sister Bay on the map.

In addition to mowing, rooftop goats can provide **milk, meat** and sometimes

Roosters are almost always banned [by local municipal codes], so focus your energy on the ladies!

Photos clockwise:

*Blushing Goat
Farm, CO*
PHOTO COURTESY OF
LEAH CAPEZIO

Terra Fata Farm, CO
PHOTO BY TRENTON BARNES,
COURTESY OF
EMILY HARTNETT

*Eagle Street Rooftop
Farm chicken coop, NY*

wool (depending on the breed), while occupying a relatively small area. Their milk can be used for drinking and producing dairy products such as cheese and yogurt, and also for making soap and other value-added products. Sheep and cows provide similar services and products as goats but are less appropriate animals for rooftops. Sheep require a larger grazing area than goats, and therefore provide less benefit per square foot of roof space. In other words, if one sheep and one goat produce equal amounts of milk, meat and wool, but a sheep requires twice as much roof space as a goat, then a goat is twice as valuable as a sheep. Cows, as ruminates, require a fully grass diet to remain healthy. Growing grass on a roof to graze cattle is impractical at best. Given the space constraints of urban roofs, goats are the most valuable rooftop herd animal.

Each city specifies which farm animals are permitted within city limits, how many can be owned at once and which ones require a permit, license or registration. A regulatory comparison chart in Chapter 6 provides a complete listing of legal farm animals in eight North American cities, as of August 2012. Be sure to check with local health codes before investing in some feathered, furry or wooly friends, as these particular regulations tend to change over time.

Grazing rooftop goats,
Al Johnson's Swedish Restaurant + Butik, WI
PHOTO BY MATT NORMAN PHOTOGRAPHY, COURTESY OF
AL JOHNSON'S SWEDISH RESTAURANT + BUTIK

beekeeping

Without a doubt, the most necessary farm animal of all is the honeybee. In her book, *The Beekeepers Lament: How One Man and Half a Billion Bees Help Feed America*, author and award-winning journalist Hannah Nordhaus reports that, in the US alone, "farmers depend on honeybees to pollinate ninety different fruits and vegetables ... nearly $15 billion worth of crops a year."[7] These astounding statistics remind us that our food system depends on our tiny winged friends, even if the food is grown on a roof. Rooftop beekeeping is on the rise, and many skyline farms maintain hives in order to ensure that their rooftop fruits and vegetables become pollinated. A by-product of these busy pollinators is of course **honey**, which makes bees even more valuable, in that they provide both a service and a product.

Well-seasoned beekeeper and founder of the Philadelphia-based company Urban Apiaries, Trey Flemming, continuously observes that **bees living in cities are generally "healthier and more productive" than rural bees,** and don't require as much supplementary nutrition during the colder months.[8] While this may seem shocking, urban bees avoid the pesticides that rural bees regularly encounter when flying through non-organic agricultural fields.

So what's needed to keep bees on a rooftop? Well, you'll need a hive to house the bees and their honeycomb; a parapet, wall or windbreak to protect the hive; a clear landing and takeoff zone in front of the hive; a water source that's shallow enough to prevent drowning; a smoker to calm the bees before opening the hive; a tool to scrape the supers inside the hive; and you'll need bees. Lots of them. The European honeybee, *Apis mellifera*, is a canopy dweller by nature, which means

Beekeeper Trey Fleming checking his rooftop hives at SHARE Food Program, Urban Apiaries, PA

they actually favor trees and rooftop hives over ground-level residences. Bees can find their way up to church steeples and high buildings, particularly if tall trees are present to act as a kind of stepladder.

In 2011, Flemming told me that the biggest obstacle to urban beekeeping in the US is "convincing people that the bees aren't going to sting you." The European honeybee is a gentle creature, and yet beekeeping is illegal within many North American cities. In New York City, for example, the Department of Health and Mental Hygiene classified the pollinators as an "animal nuisance" along with venomous insects and dangerous dogs.[9] Due to beekeeping's rising popularity coupled with increased awareness and education, keeping non-aggressive honeybees in New York City is now legal, as of 2010. Apiarists are required to register their hives and follow appropriate beekeeping practices, such as maintaining each hive and providing a constant water source. Beekeeping in Philadelphia was never illegal, but urban apiarists must take care to run their operations appropriately, so as not to deleteriously affect the legality of their well-intentioned activity. As urban and rooftop farmers begin dabbling in beekeeping, Flemming points out that "it takes a fair amount of knowledge to [tend hives] well." It's worth doing your homework on this one or, better yet, partnering with a veteran like Flemming.

Formalizing a rooftop beekeepers network could help to promote a positive image of safe and successful urban honey production, while enabling farmers to focus on crop production. Often times, a beekeeper gives a certain percentage of honey to the resident farmer or property owner in exchange for allowing hives on the premises.

holistic building design

A rooftop farm or garden may be just the garnish your building needs to stand out from the crowd. Taking things a step further, the rooftop may tie seamlessly into the structure below and become integral to the building's functioning. Holistic building design involves thinking about a building as a single functioning organism, with various integrated systems that, together, make the structure tick. Rooftop farms and gardens play perfectly into this design approach, as integrated systems often address water use (and sometimes reuse) within a building, indoor temperatures, indoor air quality, waste cycling, etc.

The stormwater diagram shown here shows how rooftop farms and gardens can absorb stormwater and direct excess flows to an underground storage tank, or cistern. The water can then be filtered, pumped up through the building and used for irrigation or flushing toilets. This simple concept reduces the amount of water entering the sewer system, while simultaneously decreasing the building's dependence on municipal water. Two birds, one stone. Water can additionally improve building performance by regulating indoor air temperatures. Since water gains and loses heat more slowly than conventional roofing materials, saturated soil on a rooftop actually insulates one- or two-story buildings from temperature fluctuations. Indoor temperatures consequently remain cooler during summer months and warmer in the winter, which leads to more digestible utility bills and a more pleasant indoor environment.

Rooftop greenhouses are another way to influence building temperature. Mohamed Hage, founder and president of a 31,000-square-foot rooftop greenhouse operation in Montreal called Lufa Farms, (see chapter 5) told me that the temperature inside his Canadian greenhouse couldn't be more pleasant, even in winter.[10] Depending on how a glass structure connects to the building below, the heat generated within a greenhouse can sometimes be harnessed and fed into adjacent stories. The sun can also be harnessed through the installation of photovoltaic (PV) panels on a rooftop. When considering a PV array, keep in mind that **solar panels and rooftop agriculture are not mutually exclusive.** More structurally sound roof areas may be perfect for growing food, while weaker areas may be just right for harnessing energy from the sun. This energy can be converted into electricity to power light bulbs and all sorts of things within a building.

Rather than getting tied up in the details of each speck of soil on your roof, think big about how your enclave can tie into, and even improve, the surrounding built environment.

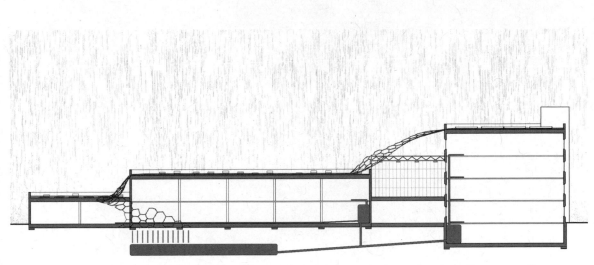

Stormwater diagram

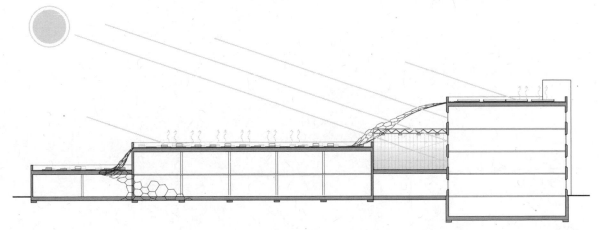

Solar diagram

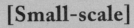

[Small-scale]

4 | Rooftop Gardens [Small-scale]

Rooftop gardens as small as a potted tomato plant and as large as a community garden dot our urban skylines. These sacred spaces seem to sprout up overnight above townhouses, apartment buildings, schools and churches. And why are they so popular? Because it's so darn easy to grow your own vegetables!

Rooftop home gardening is simple enough for anyone to try, no matter how inexperienced or squeamish you may be. All you have to do is give it a try and dig in. Here are a few examples of inspirational rooftop gardens to whet your appetite and encourage you to try it for yourself.

Tomato plants and solar panels atop a West Philadelphia row home, Sand Residence, PA

Home gardener Jay Sand and his three daughters,
Sand Residence, PA

Rooftop Gardens [Small-scale] **73**

Sand residence

Philadelphia, PA

the garden

Jay Sand and his wife Lauren decided to renovate their spacious Victorian fixer-upper from bottom to top, with one key addition: a rooftop vegetable garden.

"We wanted the fun of the house to extend up to the roof," Sand explained, when I visited his Philadelphia town house in 2012. Now the whole family spends time together outside, while working the soil and experiencing the joy of fresh food.

The garden itself, roughly 200 square feet (0.004 acre), sits on a wooden roof deck next to a solar array. Sand hired an architect to design the deck, which includes a railing and rooftop spigot. The garden is accessible through a small door in the attic, which used to be a window.

Twelve **self-watering containers** line the deck, each made from two plastic bins (one nested inside the other). An overflow hole drilled through each bin allows excess water to escape, and one flexible pipe, several inches taller than the bin, extends down to the base of each inner planter. Once a week, Sand connects a garden hose to the pipes and fills the reservoirs that are created by stacking the bins. The plants take up water as needed, and a layer of black plastic over the soil further prevents water loss by minimizing evaporation. The soil itself is nothing fancy — just a blend of coir, peat and compost. Jay was not overly concerned about the weight of the system, which allowed him to avoid the world of engineered, lightweight soils.

The garden produces enough vegetables for Sand and his family to snack on throughout the summer. Tomatoes, cucumbers, brussels sprouts, beans and mint dominated the garden during my visit, and each year the family experiments with new crops or growing methods. In 2011, they planted a melon, which grew out of control on the roof!

the gardener

Like most home gardeners, Sand has a day job. The 40-year-old Harrisburg, PA native teaches music lessons in his West Philadelphia row home. When he's not working, Sand is busy with his three daughters: Molly (eight), Lily (five) and Adaline (three). When I joined the girls up on the roof, they frolicked around the garden while describing their favorite uses for mint and explaining the importance of self-watering planters.

Toward the end of my visit, the girls shared a freshly picked rooftop cucumber. They passed around the snack, each taking a bite, until the veggie was demolished. With the realization that the cucumber was gone, little Adaline burst into tears screaming, "Daddy, I want a cucumberrr!" The tot had developed such a deep connection to fresh vegetables that only the promise of more cucumbers could console her.

secret to success

Sand did not grow up in a hands-on household, although you would never know it when meeting him. He comes across as an enthusiastic entrepreneur, always willing to get his hands dirty and experiment. His rooftop garden reflects this pursuit of raw **experimentation,** as Sand had never heard of rooftop vegetable gardening when he conceived of the home improvement project. He bought a book on ground-level container gardening, performed some **online research** and developed a planter prototype in no time.

Sand's advice to rooftop home gardeners is to "think [the project] through, but don't overdo it! Just try it." It's important to be practical about the structural limitations of your house, rooftop access, etc., but Sand's point is to move forward, test methodologies and innovate.

Sand's rooftop gardening success prompted him to organize a group called the **Philadelphia Rooftop Farm (PRooF)** in 2009. The active group promotes rooftop gardening across Philadelphia's flat roofs and provides information on its website for homeowners interested in building rooftop gardens. The group seems to fuel Sand's activist spirit and allows him to interact regularly with the community.

Lily Sand, 5, with a giant rooftop cucumber,
Sand Residence, PA

Jay Sand interview
Home Gardener + Philadelphia Rooftop Farm (PRooF) Organizer

LM How did you become interested in vegetable gardening?

JS My wife had a ground garden when we lived in West Virginia …. When we were thinking about what to do with the roof [on our Philadelphia home], I decided that [rooftop gardening] was something I'd like to try. I bought a book about container gardening … and researched online. Within a few weeks I built a prototype …. I didn't go through any failures with this particular part. It became something I enjoyed because it was **instantly satisfying**, and I, as a novice gardener, could do it. I don't have to be a specialist or have any real talent [at gardening] to have a [successful] rooftop garden.

LM What growing methods work best for you?

JS I tried the self-watering planter first, and it worked so well, I found no reason to try anything else …. Tomatoes and cucumbers are super easy and they grow really well. So far everything has worked out really well.

LM How much food does your garden produce annually?

JS We eat it so fast that it barely makes it downstairs [to the kitchen]! We cook with the greens …. It's a nice addition to our regular food intake …. We constantly eat the vegetables [raw], and actually cook with them once or twice a week. We like cooking with chard, eggplant and peppers. We use the tomatoes too.

LM Do your kids help you harvest the rooftop vegetables?

JS Oh yes! They're all about [the garden]. They invite other kids over and teach them how to garden.

LM Why did you decide to grow on your roof?

JS It was an instinct and an idea. [My wife and I] had this plan of extending the fun of the house to every square inch of space. I also **liked the idea of a green space** …. It's really pleasant to be up [in the garden], even in the summer sun. The back yard is full-shade — we can't even grow grass back there. The roof was the only place [on the property] we could grow anything.

LM What is the greatest joy of gardening on your roof?

"I don't have to be a specialist ... to have a successful rooftop garden."

"The back yard is full-shade — we can't even grow grass back there."

JS It's fun having a wholesome project with my kids. The kids get excited, and we do something that's all good. I also like the idea of doing things myself. I didn't grow up with a hands-on approach, and so I wanted to take the opportunity to do things myself.

LM Why is urban agriculture meaningful to you?

JS I love living in the city. The balance of diverse people, things to explore and learn …. We have the best of both worlds with the "country" on our roof. I felt that I didn't have to make the choice between the city and the country. I also think it's important for urban kids to have relationships with soil and food.

LM Do you end up saving money by growing your own food?

JS I don't think we do, because we don't grow a substantial enough amount to replace what we're buying. Some of the vegetables [that we grow] are pretty expensive to buy, and so we wouldn't be buying them anyway.

LM What has been your greatest challenge?

JS Finding the time at the beginning of the season …. The success of the garden involves incrementally putting more thought into it [each year] and being willing to try anything to see if it works.

LM Is rooftop gardening easy enough for anyone to try at home?

JS Naturally. You have to be smart about it if you're going to put weight on your roof. You have to think [the project] through, but don't overdo it! Just try it …. As someone who has a busy existence, I like to do things that have **many benefits** and few risks …. Farming is really difficult, gardening is pretty easy.

LM Is rooftop gardening a community activity?

JS A new neighbor is growing [with containers] on her roof …. **Food is a unifier** … and sharing it has connected people over time.

> "It's important for urban kids to have relationships with soil and food."

> "The success of the garden involves incrementally putting more thought into it [each year]."

The Sand Family enjoying their rooftop garden,
Sand Residence, PA

Alarcon residence
Philadelphia, PA

the garden

In 2012, the Philadelphia Rooftop Farm (PRooF), a local group of activists and gardeners, installed a pilot vegetable planter on top of a West Philadelphia row home. The group conceived of the project in 2009 and quickly decided to elicit design assistance from the City's Community Design Collaborative (CDC). The CDC connects teams of volunteer design, engineering and cost estimation professionals with worthy projects around the city.

"[PRooF] had an idea of what they wanted to do, but didn't know the practicality of how to get there," explained architectural designer and CDC team member Gavin Riggall, when I spoke to him after the planter was installed. In addition to Riggall, the team consisted of architects, a structural engineer and an estimator. The professionals evaluated Philadelphia's single-family residential building stock and concluded that row homes have the most potential for rooftop gardening.

The downside? The roofs of these buildings are typically not strong enough to support much weight. The CDC's planter design, therefore, takes advantage of a row home's strongest rooftop component: the party walls. These knee-high walls that separate one row home from the next can bare more weight than any other part of

the building. Suspending a planter system from one wall to the next keeps weight off of the roof itself, thereby allowing the homeowner to plant in deeper, heavier soil (allowing for higher crop diversity).

PRooF member Stephanie Alarcon volunteered the roof of her West Philadelphia home for the pilot planter installation. The only catch was that the most accessible part of her roof, a lower roof area above an addition, did not contain party walls. The group decided to install the planter directly on the waterproofing membrane, which ultimately caused a leak. They decided to remove the planter and reinstall it according to the CDCs original plan.

The planter itself was built as designed, with four plastic liners inserted into a wood frame made of pressure treated two-by-sixes. PRooF volunteers placed soda crates at the base of each liner to create a reservoir, and then installed fabric, soil, black plastic and plants. Each planter is self-watering, just like the containers at the Sand Residence. The CDC design is modular, so that several of these planters could be suspended on a single roof (weight permitting, of course).

the gardeners

While the installation was a group effort, Alarcon performs the day-to-day gardening. She fills the reservoirs with water, harvests produce and keeps a general eye on the planter. PRooF's original plan was to install planters on homes across the city, tend them with a gardening labor force and sell a portion of the produce at market. Philadelphia building codes require the installation of certain architectural elements (e.g., a headhouse, secure railings) when a rooftop is regularly accessed for commercial purposes. These requirements, compounded by associated operational liabilities, influenced PRooF to adjust its goal. Now the group seeks to inspire Philadelphians to build their own planters, while providing them with helpful resources.

secret to success

PRooF members learned a tremendous amount about rooftop gardening through their building code **research**, design development, **collaboration** with the CDC and planter construction. They also gained valuable insight into roof leaks, and the importance of **consulting** with professionals and heeding their advice in order to avoid such occurrences. Learning from mistakes is a key step to innovation and progress. Whether those mistakes are your own or someone else's, disseminating the experience and the associated "lessons learned" can be invaluable. PRooF's next

installation will build upon the information gained from the Alarcon Residence pilot project. The group is optimistic, and is eager to pursue the next planter experiment.

PRooF volunteers planting vegetables,
Alarcon Residence, PA

Volunteers installing the rooftop planter prototype,
Alarcon Residence, PA

1

Insert the plastic liner into the wooden frame. Metal straps, suspended from the wooden frame, will support the liner. The liner will fit snuggly, and the top should be flush with the top of the frame.

2

Install a reservoir layer at the base of the system. The material's void space will hold water, while its rigid structure will support the weight of the media above. This planter contains plastic soda crates, but a conventional green roofing reservoir sheet or even a sheet drain will do.

3

Lay landscape fabric (or separation fabric, as it's called in the green roof industry) across the reservoir layer and up the sides of the liner.

4

Lay a second layer of separation fabric across the planter, perpendicular to the first piece. Covering the inner corners of the liner with fabric during this step will prevent soil from entering the reservoir and clogging the system.

5

Cut a hole through both layers of separation fabric to accommodate a metal or ultra-violet stabilized plastic pipe. This pipe should reach the base of the system, and will be used to fill the reservoir with water. The pipe should be wide enough to accept a garden hose nozzle.

6

Fill the planter just below the brim with a planting soil of your choice. Covering the soil with black plastic will slow weed growth and evaporation, but will also prevent rainwater from entering the system — the plastic is optional.

Gavin Riggall interview

North Street Design, Partner + CDC Team Member

LM What was your personal interest in the project?

GR The mission of my [design] company at the time was to develop products that addressed the **interface between architecture and the outdoors.** Utilizing stormwater and space, and improving quality of life were priorities.

LM At what stage was the project when you began working on the planter design?

GR The conceptual stage. **[PRooF] had an idea of what they wanted to do, but didn't know the practicality of how to get there.** That's why they went to the CDC. We had a structural engineer, architect and estimator all on hand to provide their expertise.

LM Why did the CDC team decide to design a suspended planter?

GR We can't place this amount of weight directly on the roof of an old building, specifically on a row home in Philly. Some of these homes are over 100 years old …. We had to figure out a way to transfer the loads of these planters, which can be thousands of pounds, to the strongest part of the roof. We decided to bridge between the party walls to transfer the load without compromising the structure of the existing building.

LM What is the longest span the planter can reach with the wood-framing design?

GR We looked at the standard 14 to 16 feet [the typical Philadelphia row home width]. The price of lumber really starts to go up astronomically the longer [the lumber] gets.

LM What was your greatest design challenge?

GR Figuring out **how to get up there in a legal, safe way** to farm on a roof in the city of Philadelphia.

LM What potential does the planter have for spreading to other rooftops?

GR If [PRoof] can nail down ease of installation and the structural considerations of the existing housing stock, [the planter prototype] could definitely be deployed across the city, economically and efficiently …. The amount of unused roof space across the city is staggering, and so **it would be a shame not to utilize [this simple technology]** for growing food throughout the city.

> "We decided to bridge the party walls to transfer the load."

graze the roof
San Francisco, CA

the garden

Seven stories up in the heart of downtown San Francisco, Graze the Roof occupies 900 square feet (0.02 acre) of Glide Memorial Church's roof. The garden's community-driven mission emphasizes education, demonstration and empowerment, thereby serving as an invaluable asset to visitors, many of whom are low-income children from nearby neighborhoods.

Activist and visionary Maya Donelson established the garden in 2008 with the help of a $10,000 grant. "The dream was to transform an underutilized surface ... into a vibrant landscape of food, community and education," explained Graze the Roof project manager Lindsey Goldberg, with whom I corresponded in 2012. The garden now operates in partnership with the Glide Foundation, a charitable organization founded by Glide Memorial Church, which fully funds the project. Director of Facilities Terry Zukoski runs Graze the Roof, with project managers Goldberg and Nikolaus Dyer at his wings. According to Goldberg, the small staff welcomes 250 to 450 volunteers and 300 to 500 visitors to the garden each year. Community work parties, monthly workshops and an after-school program, as well as the promise of fresh vegetables, help to reel people in. Volunteers are encouraged to enjoy the fruits of their labor, while the remaining produce is donated to Glide's educational programs and soup kitchen in the building below.

The gardeners cultivate vegetables, herbs and flowers within raised planters (made from recycled milk crates), five gallon buckets, hanging pots and burlap sacks hung to the surrounding fence line. Concerns about load restrictions lead the garden's founders to consult a structural engineer prior to construction, and as a result, lightweight soil and thin soil systems were installed. Goldberg reported that the lightweight soil consists of potting soil, perlite, compost, worm castings and coconut coir, with small amounts of mulch and sand. Drip irrigations and an onsite vermiculture (worm composting) operation help to keep crops happy as they bask in the sun.

the gardeners

Goldberg and Dyer keep busy by managing production, organizing volunteer activities and educating visitors about gardening and nutrition. The California natives, in their late twenties and early thirties, have been involved with food production for the past three and ten years, respectively. Before growing food, Goldberg studied language, culture and geography at California Polytechnic State University, and then earned a Masters of

Project Managers + Garden Educators
Nikolaus Dyer (left) and Lindsey Goldberg (right),
Graze the Roof, CA
PHOTO BY MICHAEL I. MANDEL, COURTESY OF GRAZE THE ROOF

Education in Seattle, with an emphasis on environmental and arts education. While I have never met Goldberg, her emails suggest an unmistakable energetic demeanor, not uncommon among young activists and educators. When I asked how she first became interested in urban agriculture, Goldberg replied, "I wanted to live in San Francisco and continue doing work that I love: connecting people to the natural world, building community and inspiring connection to place. Urban agriculture was the perfect landscape for [fulfilling] both my professional aspirations and personal passions. I *love* food and gardening, cooking, eating and building community with people from all walks of life!"

secret to success

The energy and enthusiasm of Graze the Roof's staff, combined with the garden's **socially minded mission** and its steady stream of **volunteers**, provide a recipe for success. The Glide Foundation's financial backing, of course, furthers this success. For mission-driven gardens like Graze the Roof, associations with **parent organizations**, angel investors or other beneficiaries takes the pressure off fundraising. Time can be spent instead with organizing, coordinating and teaching, learning and gardening! Lastly, the public nature of the building below further fosters the garden's success. The highly trafficked building, with easy rooftop access, means that more people can visit the roof than in a private, less accessible building.

Lindsey Goldberg interview

Graze the Roof, Project Manager + Garden Educator

LM What is Graze the Roof's mission?

LG To educate people about how to grow food in the urban environment, empower people to make **healthy food choices** through understanding our food system and to demonstrate low-cost container gardening.

LM What is the garden's association with the Glide Foundation?

LG Graze the Roof is the rooftop garden atop the Glide Foundation Building. We are [fully] funded through the Foundation.

LM Who founded Graze the Roof, and why was it founded?

LG Graze the Roof was founded in 2008 by a team of visionary volunteers, led by a young woman, Maya Donelson. The dream was to transform an underutilized surface ... into a vibrant landscape of food, community and education.

LM Do you consider Graze the Roof to be a community garden?

LG Absolutely! Graze the Roof is supported by the volunteers who show up each week to help with the development of the garden. These volunteers take a vested interest in the success of the space, and benefit from the abundant harvest we are generating!

LM What crops are you growing right now?

LG Lettuce, radish, mizuna, kale, red mustard (ruby streak), tomatoes, basil, leeks, red onion, fava beans, rainbow chard, violas, nasturtiums, scarlet runner beans, bush beans, pole beans, carrots, mint, oregano, dill and lots of succulents. [We've had the most success with] lettuce, tomatoes, basil, beans and strawberries.

LM Where is the food distributed, sold or consumed? By whom is it consumed?

LG We harvest ... **5 to 15 pounds of food each week** and integrate that into the GLIDE kitchen, which distributes 3,000 meals each day. Volunteers also take home a big bag of produce at the end of each workday. The children served by the educational programs also benefit — they love snacking on strawberries and munching raw kale!

"The dream was to transform an underutilized surface ... into a vibrant landscape of food, community and education."

LM In what ways do the garden and its programming build community?

LG [We host] **weekly work parties,** [which provide community members the opportunity to] work together on an array of different projects connected to the development of the garden.

LM In what ways do the garden and its programming educate youth?

LG [Through the Glide Foundation], we have a partnership with the Family Youth and Childcare Center, which serves 75 youth [in San Francisco's] Tenderloin neighborhood. These youth participate in an **after-school program** ... at the garden, three days a week.

LM Does Graze the Roof promote healthy eating and educate kids about nutrition?

"Cooking and nutrition are an integral part of our programming."

LG Yes! Cooking and nutrition are an integral part of our programming.

LM What rooftop events has the garden hosted?

LG We host **monthly workshops** related to urban agriculture, including: planning your spring garden, [worm composting], natural building, fermentation, irrigation ... and [edible green walls].

LM What has been the garden's greatest success?

LG The day-to-day joy, creativity, healthy challenges of growing an incredible amount of food on a rooftop in the heart of a rough neighborhood. Witnessing people's awe and respect for this project and **cultivating a ritual around a weekly harvest** that gets donated to the GLIDE kitchen.

LM What have been the garden's greatest obstacles?

LG Funding. Our budget is minimal, and it definitely slows the garden's development.

LM Would you consider expanding to other rooftops?

LG Graze the Roof keeps us really busy, but I'd love to collaborate with other gardeners to activate more rooftops in the city.

Vegetables and herbs grown in salvaged containers,
Graze the Roof, CA

PHOTO BY MICHAEL I. MANDEL, COURTESY OF GRAZE THE ROOF

garden checklist

So you want to grow vegetables on your roof? Fabulous. Now how the heck do you get started? The Garden Checklist provides step-by-step guidance to maximize your project's success, before ever picking up a hand spade. Whether you're an experienced or first-time gardener, the Garden Checklist will introduce you to the most essential rooftop considerations, from railing requirements to irrigation techniques. Following these easy steps will help you think through your project before getting started, so that you, your family or your friends can enjoy trouble-free construction and gardening!

Nikolaus Dyer tending to alpine strawberries, Graze the Roof, CA

PHOTO BY MICHAEL I. MANDEL, COURTESY OF GRAZE THE ROOF

☑ **Building codes**

    ☐ Rooftop access meets local building code requirements

    ☐ Setback requirements reviewed and understood

    ☐ Railing system meets local building code requirements

☑ **The building below**

    ☐ Structural capacity approved by engineer

☑ **Up on the roof**

    ☐ Roof receives full sun and little wind

    ☐ Water hookup available and fully functioning

    ☐ Planter placement understood

☑ **The nitty-gritty**

    ☐ Containers/raised beds selected to fit your roof

    ☐ Soil mixes researched

    ☐ Planting plan developed

    ☐ Irrigation methods researched

    ☐ Gardening schedule considered

1. building codes

rooftop access

Access and **safety** are two of the most important considerations of rooftop gardening. Local building codes are designed to promote public safety in and around your building, including up on the roof. Building codes vary by city, and so step one in planning your rooftop garden is to research exactly what restrictions apply to your project. In Philadelphia, where I live, building codes specify that you cannot access a roof regularly (more than a few times per year) by ladder or pull-down staircase. These precarious methods of access are particularly dangerous if you're carrying something awkward like compost or seedlings up to the garden.

The safest **legal** way to access a roof regularly in Philadelphia is through the door of a taller building story or a **headhouse** (a rooftop vestibule in which a staircase from the floor below exits onto the roof). Installing a headhouse can be prohibitively expensive for most homeowners, and so be sure to confirm that access is possible and within your budget. Check with local building codes and your apartment building manager to see if rooftop access is permissible via fire escape. Rooftop gardener Jay Sand, from the Sand Residence, replaced an attic window with a door, in order to improve rooftop access and avoid having to crawl through the window.

setback

In addition to promoting public safety, building codes are intended to preserve and promote certain exterior aesthetics. Particularly in historic neighborhoods, these codes may address what rooftop elements (or additions) are visible from the street. This type of building code specifies a setback, or specific no-build zone, from the street-side edge of the roof. The purpose of this type of **setback** is to "hide" unattractive structures from neighbors and pedestrians; the result may be that your garden space is quickly reduced to nil. If your project is subject to this building code, identify the restriction and its effect *before* you begin construction, and apply for a variance through the local building commission. The worst scenario would be to build your garden, incur a hefty fine and have to tear everything down.

railing system

Similarly, building codes dictate what protective infrastructure must be installed around the **roof's perimeter** to enable regular legal access. Apartment buildings may contain a perimeter knee wall called a "parapet." If the **parapet** is at least 48 inches high, then rooftop access is permissible by most building codes. If the roof contains no second form of egress (like a fire escape), then building codes may restrict how far your garden can extend from the door or headhouse. Again, this is for your safety.

If the parapet is not tall enough or does not exist at all (in the case of many row homes), then building codes may dictate that a secure **railing** be installed. The railing itself must meet code, which means that it has to be able to withstand a specified amount of horizontal force. Bolting into the roof deck or parapet is expensive and can lead to leaks. The best option is to secure the railing to another structure, such as the perimeter of a deck. This method is particularly inexpensive and effective for a small home installation. If young children will visit the garden, then make sure that the fence openings are small enough to prevent kids from climbing through — consider installing wood lattice fencing on the inboard side of the structural fence. Additionally, the fence should not enable young children to climb up and over — consider vertical slats.

Mixing lightweight soil by hand, Alarcon Residence, PA

Secure perimeter fence fastened to decking,
Sand Residence, PA

2. the building below

Roofs are like glass tables: they can easily support a heavy book, but you might think twice before standing on top of one. Like tables, certain types of roofs are meant to bear more weight than others. The amount of weight that a roof can support depends on the roof deck material, its core strength, the building's beam and/or column spacing, beam and column strength and the design strength of the structure's foundation. In the case of older buildings, the level of deterioration of wooden and concrete components infinitely affects the roof's load capacity. As a sweeping generalization, older industrial buildings with steel (or steel-enforced) beams and columns possess some of the strongest roofs, while newly constructed row homes generally contain some of the weakest. To complicate matters further, small buildings like row homes often have zones throughout the roof that are weaker than others. This is complicated!

With all of these compounding variables, how are you supposed to know how much weight your roof can sustain? You're not. You should never assume how much weight your roof can support, even when looking at the building's blueprints. Buildings are often not built as designed, and even if they are, they can weaken over time. *Always* consult a **licensed structural engineer.** An official structural analysis is the only way to ensure that your roof will not collapse under the weight of heavy soil. Don't skimp on this one — it's important.

Container placement reflecting load constraints,
Graze the Roof, CA

Photo by Michael I. Mandel, courtesy of Graze the Roof

3. up on the roof

microclimate

Step three involves making sure that your roof's microclimate is suitable for vegetable production. Ideally, the garden space should receive **full sun** and **little wind.** The former requires a south-facing roof, unobstructed by the shade of surrounding structures. Buildings in most urban residential neighborhoods cannot exceed a specified height (without a variance), which means that even if your garden is north of a vacant lot, you won't risk a taller building being erected, which could block your garden's sunlight. Wind is generally a non-issue in low stature neighborhoods, particularly if your garden contains a perimeter fence. Gusts above apartment buildings can be a different story altogether. Wind blocks, such as strategically placed trelliswork, metal wind screens, low evergreen trees or even mechanical units, can help dissipate wind gusts, thereby preventing planters from capsizing. Be sure to visit your apartment building's roof during the early stages of planning, so that you know what to expect. In fact, visit several times, as some days are windier than others!

water hookup

How will you water your garden? Is it small enough to use a **watering can,** or do you need a hose? Is there a **spigot** on the roof? Does it work? These are questions that you should ask yourself before getting started.

In the absence of convenient water access, consider installing a **rain barrel** to capture stormwater runoff. Keep in mind, though, that rain barrels are *heavy*. When full, a 55-gallon rain barrel will weigh over 450 pounds, so verify with a licensed structural engineer that your roof can sustain the extreme "point load." Also carefully evaluate the condition of the surface that the water will flow off before reaching the rain barrel. Deteriorated roofing membranes can release carcinogens and other harmful chemicals into runoff, which will contaminate certain food crops when used for irrigation. Some cities do not permit the use of rain barrels for irrigating food crops, just for this reason. Please, proceed with caution and be safe!

When full, a 55-gallon rain barrel will weigh over 450 pounds.

container placement

As mentioned previously, a roof's strength may vary by location across the roof. On large buildings with columns, the most weight-bearing areas of the roof are typically directly above each column. On most row homes, the roof areas along the party walls (knee-high walls separating one home from the next) are typically strongest. Keep this rule of thumb in mind when determining your garden layout, and what types of containers or raised beds you'd like to incorporate. Many rooftop gardens place heavier planters around the perimeter and lightweight containers within the roof's interior or directly on the fence line.

A roof's strength may vary by location across the roof.

4. the nitty-gritty

planter selection

If you recall from Chapter 3, rooftop planting strategies typically fall into four categories: container gardening, raised bed production, row farming and hydroponic production. Most rooftop home gardeners choose to grow food in containers and raised beds, of all different materials and styles. With an endless variety of planter options, your goal should be to **determine what will best suit your garden space and your lifestyle.**

First, think about how much growing space (actual soil area) you will need. Are you planting a small herb garden for yourself? Will you grow vegetables to feed your family of four? Do you plan on growing as much as possible to share with neighbors and donate to a food bank? Evaluating how many planters you will need will help you establish a realistic layout and purchasing plan. Also, keep in mind that **most gardens are built** *incrementally*. Try starting with three or four containers. If everything goes well, then go ahead and expand your garden! The bottom line is to map out your long-term vision during the initial stages of planning, so that you can easily expand your garden without too much hassle and extraneous lifting along the way.

If you plan on experimenting first with containers, then search online and visit local garden centers to see what's available. Select shallow containers for greens and herbs, and deeper containers for crops like tomatoes, eggplant and brussels sprouts. If you're feeling adventurous, then try assembling your own containers out of salvaged or recycled materials. This strategy is particularly popular with home gardeners who favor self-watering planter bins. Regardless of what type of containers you select, keep in mind that soil in ceramic pots dries out more quickly than that in plastic or metal.

If raised beds are on the docket, then be sure a licensed structural engineer has approved your layout plan. Raised beds will prove less mobile than containers, but your plants will perform more robustly as the temperature shifts throughout the growing season. You may decide that placing the beds around the roof's perimeter makes the most sense, thereby leaving the garden's center open for containers and social space (such as a table and chairs).

soil mix

Now that you've thought about planters (perhaps at nauseam), what type of soil will you put in them? If you're new to gardening and simply testing out a few small containers, then try using an off-the-shelf garden soil, available at any garden center or hardware store. For larger projects, particularly when weight is of concern, most home gardeners mix their own lightweight soil. Everyone's "home brew" is slightly unique, and many gardeners

Volunteers blending a lightweight soil mix with peat,
sand, perlite and garden soil,
Alarcon Residence, PA

choose to experiment and modify their mix over time. During the Alarcon Residence rooftop garden, volunteers opted to vary the soil mix in each planting area in order to evaluate which performed most effectively. The mixes all involved different quantities of peat (for water retention), sand (for drainage), vermiculite (to prevent soil compaction) and off-the-shelf garden soil (to act as a base). Many gardeners also mix compost into their soil to provide the plants with nutrients.

For large installations involving numerous raised beds or even a full-fledged rooftop community garden, some gardeners choose to purchase lightweight green roof media from one of the many media blenders located throughout the US. Blenders such as Skyland USA (with a national network of certified distributors) offer an "agricultural media," while other blenders like Stancills, Inc. (in MD) offer more custom solutions. These mixes are each engineered with a specific pore space, water-holding capacity, saturated weight, organic content (usually 6% to 10%) and other properties. Commercial media comes fully blended, which saves you from having to mix large quantities by hand. The organic content is notably lower than that of typical agricultural soil, which means that you will need to amend the media regularly with nutrients. Using chemical fertilizers will degrade your media over time, and runoff from the planters will enter the sewer system and ultimately

pollute our streams and rivers. To keep your media light and non-compacted over time, try feeding your plants with organic supplements such as compost tea, fish meal or bat guano. If weight is less of an issue, add compost or decomposed horse manure.

planting plan

Developing a detailed planting plan in advance of a farming season is critical, but it is generally not a necessary step for most gardens. Instead, think back to your overall goals for the garden (raising herbs, feeding many people, etc.) and decide what types of vegetables and herbs you would enjoy. When it comes time to planting, keep in mind that all crops should receive as much sun as possible (don't let tall crops block the sun for shorter crops), and be sure to position the plants to allow for accessible harvesting. More advanced gardeners may consider planting strategies such as companion planting and other principles of permaculture and square-foot gardening.

irrigation

Rooftop containers are typically watered either by hand (with a hose or watering can) or through a **self-watering** system (with a reservoir below the soil). Raised beds can also be watered by hand, or alternatively with drip irrigation, which involves installing perforated **drip lines** that slowly release water. Most raised bed drip lines are installed on the surface of the soil, but try burying the lines by one to two inches if you're feeling adventurous. **Sub-surface irrigation** is a common strategy within the green roof industry for reducing surface water loss. Burying the lines in the root zone may also encourage deeper root growth, which could make your plants hardier during hot spells.

Exposure to rooftop wind and sun causes planting soil to dry out more quickly than on the ground, and planting in containers rather than a contiguous planting bed exacerbates this drying. No matter what type of planters and irrigation you select, plan on watering your crops more frequently than you would on the ground. If you'd like to minimize the time that you'll spend on this chore, then seriously consider self-watering planters.

gardening schedule

If your family owned a dog while you were growing up, you may recall the pre-purchase pleading, the initial post-purchase excitement and then the inevitable disenchantment over feeding regiments and early morning walks. "Dad, do I have to take the dog out *again*?" Gardens require a similar level of dedication. Every garden needs a steward. Maybe you will be in charge of watering, tending to the plants and harvesting. Maybe these activities will be your kids' responsibility. Determining who will assume primary care of the garden is essential in promoting healthy plants and a positive experience for everyone involved. If your household, group or rooftop community garden members would like to share the stewardship role, then consider developing a gardening or watering schedule.

Every garden needs a steward.

checklist, check!

Now that you're acquainted with a range of planning consideration, try filling out the **Garden Checklist** to see if you're ready to get started! How did you rank? Did you check off every box? If not, what can you do to complete the list? Will more extensive research be sufficient? Can a friend help you overcome the trickier steps? Gardening is a community activity within cultures around the world, as collaboration breeds knowledge. If you're stuck on something, don't be shy about asking for help.

Talk to your friends, your neighbors, even your kids! Sometimes thinking outside the box can spur the greatest leaps toward innovation.

While filling out the **Garden Checklist**, remember that each rooftop garden is unique in wondrous ways. Steps that make or break your neighbor's project may not dictate the success of yours. With the helpful tips from this chapter, go ahead and get your hands dirty — what are you waiting for?!

- [ ] **Building codes**
  - [ ] Rooftop access meets local building code requirements
  - [ ] Setback requirements reviewed and understood
  - [ ] Railing system meets local building code requirements

- [ ] **The building below**
  - [ ] Structural capacity approved by engineer

- [ ] **Up on the roof**
  - [ ] Roof receives full sun and little wind
  - [ ] Water hookup available and fully functioning
  - [ ] Planter placement understood

- [ ] **The nitty-gritty**
  - [ ] Containers/raised beds selected to fit your roof
  - [ ] Soil mixes researched
  - [ ] Planting plan developed
  - [ ] Irrigation methods researched
  - [ ] Gardening schedule considered

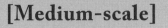

5 | Rooftop Farms [Medium-scale]

Urban rooftop farms increasingly feed and inspire people. They are powerful places. The men and women behind these iconic skyline landscapes are as unique as the farms themselves, and each brings something distinctive to the table. This chapter highlights inspirational rooftop farms and farmers from all corners of Canada and the United States. Eagle Street Rooftop Farm, Uncommon Ground, Urban Apiaries and Lufa Farms take the stage, as farming techniques, yields and personalities are considered.

Rooftop volunteers and visitors, Eagle Street Rooftop Farm, NY
PHOTO BY ISAIAH KING

Farmer Annie Novak with the Manhattan skyline beyond,
Eagle Street Rooftop Farm, NY

Rooftop Farms [Medium-scale] 109

eagle street rooftop farm

Brooklyn, NY

the farm

Iconically perched atop a warehouse building in north Brooklyn, Eagle Street Rooftop Farm looks out over the East River and Manhattan's downtown skyline. As New York City's very first rooftop row farm, the 6,000-square-foot (0.14 acre) emerald mecca supports vegetables, herbs, flowers, chickens and honeybees, all grown using organic techniques. Crops are grown in 16 three-to-four-foot-wide mounded planting beds, separated from one another by narrow mulch paths. Surface drip irrigation, and various organic topdressings, including compost, support healthy plant growth.

The building itself, a former bagel factory, is owned by the Brooklyn-based sound stage company Broadway Stages, which regularly contributes to local community and economic development projects. Broadway Stages and a New York-based green roof design-build company called Goode Green conceived of the radical project and completed construction in 2009. Farmers Annie Novak and Ben Flanner co-managed agricultural operations during the spring 2009 season, after which Novak assumed the farm manager position independently. She is supported by a market manager, farm-to-chef liaison and farm education coordinator, as well as seasonal apprentices and abundant volunteers. The for-profit farm sells produce, eggs, honey and value-added products through its Community Supported Agriculture (CSA) program, onsite farmers' market and to local restaurants and markets.

I had the pleasure of visiting Eagle Street Rooftop Farm in 2011 and witnessed for myself what all the hype is about. The farm was busily occupied by an Australian film crew (shooting a documentary segment), a German journalist (writing a story about the farm), a Jamaican insect farmer (who supplies insect-based chicken feed internationally), an aquaponics practitioner (from Growing Power, in the Midwest), volunteers, casual onlookers and myself (the aspiring author). With so much to look at, my attention flittered between Novak, colorful crop rows, the Manhattan skyline and the dissipating cloud cover — after all, I did have a photo shoot to conduct.

While the farm's layout and productivity were impressive, the space was smaller than I had previously imagined. I asked Novak about her desire to expand to other rooftops around the city, and she responded that "the need [for rooftop farming] is unbelievable .... There should be a thousand farms like Eagle Street [Rooftop Farm] in New York City."[1]

the farmer

Novak, a Chicago native and lifelong vegetarian, caught the urban agriculture bug as an undergraduate at Sarah Lawrence College. The love affair began while working with chocolate farmers in Ghana, West Africa. Since then, her passion for learning about sustainable agricultural practices has led her to Burkina Faso, Togo, Benin, Turkey, Peru, Bolivia, Argentina, Fiji, New Zealand, the Cook Islands, Alaska and Tanzania. When she is not traveling or training for a marathon (she has run seven), Novak directs Growing Chefs, a "field-to-fork" education program and co-manages the Ruth Rea Howell Family Garden at the New York Botanical Garden. On top of everything, she is integral to the daily functioning of Eagle Street Rooftop Farm.

Up on the roof, Novak gracefully juggles the responsibilities of food production, managing volunteers, answering to the media, marketing and, of course, what crops to plant next. She masters all of these tasks, always with a smile on her face. When speaking with Novak on the roof, she explained that "food connects to everything and everyone. There's nothing we do here [at Eagle Street Rooftop Farm] that doesn't have to do with the larger picture of food education, nutrition, and the environment." With her poise, generosity, articulate nature and ability

Rooftop hot pepper harvest, Eagle Street Rooftop Farm, NY

to connect people to one another, it is no wonder that Novak is a leader in the rooftop agriculture movement and a role model for budding urban farmers and activists.

secret to success

Eagle Street Rooftop Farm enjoys unparalleled success among its peers in the boutique rooftop farm world. As the **first of its kind,** the farm's novelty appeal attracts volunteers and visitors from around the globe. They arrive in droves, soak up the scene (and the breathtaking backdrop), leave with satisfied expressions and broadcast their experience to friends and colleagues. The farm's approachable **all-star staff** further contributes to this positive visitor experience. Novak in particular emits a certain magnetic appeal, and her willingness to answer questions leaves visitors and media personnel with a sense that they experienced a true taste of the farm. And it tastes *good*.

The farmer's willingness to occupy the media spotlight also contributes to the rooftop farm's success. Relative to its peers, Eagle Street Rooftop Farm excels at **marketing** and **public relations.** The well-publicized farm has garnered attention on National Public Radio, CNN, CBS Evening News, the Discovery Channel, Martha Stewart TV, the Cooking Channel, Outside Magazine, *New York Magazine*, *Edible Brooklyn* and the *New York Times*, just to name a few. The farm's publicity manager is instrumental in coordinating media visits and interviews, thereby increasing accessibility for filmmakers, journalists and authors.

Of course, the farm would not exist in the first place were it not for the generous support of Gina and Tony Argento, of Broadway Stages. The siblings' **financial backing** turned a simple idea into an actual space that feeds and inspires people. While Broadway Stages owns Eagle Street Rooftop Farm, **multiple distribution channels** allow the farm to earn its own revenue. The fact that these sales outlets (restaurants, markets, etc.) are diversified increases the potential for financial stability. If one outlet were to disengage, the others could theoretically carry the weight. In terms of complete self-sufficiency, the farm cannot rely on acreage alone. While other rooftop farms are establishing networked rooftops to achieve a scale that no single roof can provide, Eagle Street Rooftop Farm remains content above a single building. The farm does not crave monopoly acreage, and in fact, has developed innovative sales and marketing strategies to get the most out of its confined growing area through selling value-added products.

Lastly, Eagle Street Rooftop Farm's **inspirational quality** furthers the farm's success. The skyline landscape stimulates curiosity, inflates the imagination and oils the gears of innovation. If you can grow fresh, delicious food on a roof in New York City, anything is possible.

Rooftop basil harvest, Eagle Street Rooftop Farm, NY

Annie Novak interview

Eagle Street Rooftop Farm, Cofounder + Farmer

LM How long have you been growing food?

AN I started working in agriculture as an undergraduate. I focused on chocolate agriculture because it seemed people were more interested and persuaded to care about the environment and farmers' rights when given a good food, like chocolate, to tie the issues to.

LM Why are you working at this rooftop farm?

AN **Food connects to everything and everyone.** There's nothing we do here [at Eagle Street Rooftop Farm] that doesn't have to do with the larger picture of food education, nutrition and the environment.

LM What is your long-term vision for the farm?

AN There should be a thousand farms like Eagle Street [Rooftop Farm] in New York City. They should be functional green roofs. They should provide **nutritious food.** They should be filled with **passionate people.**

LM How was this specific roof selected for farming?

AN Broadway Stages chose this location, and Goode Green's [structural engineering consultant] determined it to be structurally sound.

LM What is your relationship like with the farm's downstairs neighbors?

AN The building is a television and movie production space …. So far [the relationship has been] only positive …. A lot of actors come up to check out the view.

LM Is your farm an ideal size?

AN New York needs more green roofs. So to say that this isn't the right size excludes the fact that **New York City needs more green.** We grow to suit our space — high-intensity crop rotation, lots of high-value crops.

LM Would you consider expanding to other roofs?

AN **The need [for rooftop farming] is unbelievable.** It's just a matter of finding the right **chemistry between building owner and farmer,** and identifying structurally ready rooftops.

> "There should be a thousand farms like Eagle Street [Rooftop Farm] in New York City."

LM Do you think that scale can be achieved through replicability?

AN The practical way to achieve scale is to copy your model and expand it to other roofs …. It's the same on the ground [with traditional agriculture].

LM How many CSA members do you have at present?

AN We have an Upstate [New York] farmer that we partner with. Our contribution to the CSA is to bring in **value-added** things that he doesn't offer, such as honey and eggs.

LM Can you describe your growing media?

AN It's a Rooflite [brand] growing media of about 40% compost and 60% shale and clay particulate …. We topdress with a variety of composts — aged duck manure, seabird guano and our own compost made at the farm.

LM Do you think that rooftop agriculture is viable in New York City?

AN Of course it's viable …. It's a matter of **managing labor costs** and plants that work. The City also needs to improve policy. People, plants and policy are the three Ps of successful rooftop agriculture.

LM Will your farm become an **integral facet** of the local urban food system?

AN We are an integral facet of the food system! We're a thriving, local, organic food producer.

LM Is there anything that you would like to add?

AN **Urban agriculture is not to the exclusion of peri-urban or rural agriculture** …. Our work in the city reinforces the need to support and grow the local food system all around us. As an urban site, that "educated consumer" connection is one of the most important benefits to our customers. [Additionally,] as a rooftop farm, we are using a space that would otherwise not be used, and we are managing stormwater.

> "The practical way to achieve scale is to copy your model and expand it to other roofs."

> "People, plants and policy are the three Ps of successful rooftop agriculture."

Farmer Dave Snyder amidst his organic vegetables,
Uncommon Ground, IL

uncommon ground
Chicago, IL

the farm

Chicago foodies can't resist a restaurant that harvests food from its own roof. Uncommon Ground, a restaurant on the north side of town, houses the nation's very first certified organic rooftop farm. The farm was founded on the restaurant's Edgewater location in 2007, and became certified by the Midwest Organic Services Association (MOSA) in October 2008.

The 2,500-square-foot (0.06 acre) rooftop contains an intricate patchwork of raised beds, containers and gathering space, with a lower roof area reserved for beekeeping. Planters around the farm's perimeter are actually integrated into the roof's railing system — a clever solution for meeting code while maximizing growing space. The planting beds are filled with 12 inches of organic planting soil, rich in organic content and water-retention additives. Annual compost inoculations ensure that the beds are ripe with microorganisms, which help the plants absorb nutrients. Another essential resource, water, is dished out by irrigation drip lines throughout the beds.

I visited Uncommon Ground several times in 2012 and was delighted by the farm's spring bounty. Radishes, mustard greens and lettuces were out in full force, as were blossoming chives, mint and spring peas. Before diving into a fiddlehead and asparagus salad down below in the restaurant, I had the pleasure of meeting Dave Snyder,[2] Uncommon Ground's full-time farmer. As Snyder and I chatted, he sat on a bar stool beside a pile of papers ballasted with Felco hand shears. When discussing Uncommon Ground's decision to go organic, Snyder explained that "we use our farm as an education and outreach tool ... to get people more aware of the food they're eating."[3] One benefit of the farm's organic certification is that it opens the door to conversations about organics. If you can encourage people to talk about organic vegetables, then they're more likely to try them and support the cause.

the farmer

Snyder's official title, Rooftop Farm Director, is as comprehensive as it sounds. In addition to crop planning and tending to seedlings, Snyder spends almost half of his time tending to people. Whether it's leading rooftop farm tours or training interns, coordinating with the chef or meeting with partner organizations, Snyder is busy as a bumblebee. He seems to thrive on this level of responsibility, while displaying a clear-cut passion for environmental initiatives and community activism. These interests come as no surprise for someone who lived in Seattle for 20 years.

Snyder's family kept a small side garden during his childhood, but it wasn't until after graduate school that Snyder discovered his passion for urban agriculture. Shortly after moving to Chicago in 2008, he started growing food at a community garden near his apartment. Almost all of his agricultural knowledge comes from this garden. Now, at 33, Snyder manages the Uncommon Ground farm and runs a non-profit orchard program called the Chicago Rarities Orchard Project (CROP).

secret to success

The success of Uncommon Ground's farm depends hugely on the **relationship between the farmer and the chefs**. Snyder and the head chef work closely to coordinate menu items throughout the growing season. "I have to make sure that everything I pull off the roof the chef will use," says Snyder. Part of this coordination involves growing **specialty crops** that the chef can't source elsewhere. In 2011, Snyder grew 37 crop varieties, including rare tomato breeds like the Purple Calabash. These ingredients and their unique flavors draw crowds from around the country.

The farm's pulse relies completely upon its partnerships with Uncommon Ground's two restaurant locations. The farm provides fresh produce and unlimited marketing potential, and in exchange, the restaurant pays for the farm's labor costs and ensures full demand for each and every leaf of lettuce that is produced.

Cedar planters in metal frames,
Uncommon Ground, IL

Chef Patch Carroll selecting vegetables (above)
for the restaurant (pg. 123),
Uncommon Ground, IL

Dave Snyder interview
Uncommon Ground, Rooftop Farm Director

LM Why did the farm at Uncommon Ground decide to pursue organic certification?

DS The idea was that Uncommon Ground is a restaurant that is ecologically minded. Our decisions are reflected in our food, and so we're always trying to use local organic [ingredients] …. **We thought that, if we demanded organic food from other farmers, we should be consistent with our own practices.**

LM Has the farm itself attracted business to the restaurant below?

DS Absolutely. It's a really significant PR tool …. I've heard from multiple customers that they like coming to the restaurant because the farm is so cool.

LM Has the organic certification attracted business?

DS We use our farm as an education and outreach tool … to get people more aware of the food they're eating. **Being certified organic means that we can talk about the differences between organic and non-organic food.**

LM Does the restaurant's chef request specific crops, or does your own crop selection inform the menu?

DS I keep track of my best estimates of what I'll be harvesting off the roof every week …. That provides the chef with an estimate of what will be [available at any given time]. We do coordinate. For example, if the chef wants more unusual herbs that he can't source anywhere else, I'll grow them for the menu.

LM What percentage of the restaurant's ingredients come from the roof?

DS A very, very small percentage, even during the high season. Our total growing area is a little over one hundredth of an acre…. There are **certain crops that we can entirely source from the roof, such as mint** …. Everything that we grow on the roof goes to the restaurant below, and occasionally it goes to the other location.

LM What does a day on the farm at Uncommon Ground usually involve?

"We use our farm as an education and outreach tool."

"Everything that we grow on the roof goes to the restaurant below."

DS Plant maintenance, horticultural stuff. Whether it's harvesting, pest control, weeding, I also give tours of the rooftop. There are also always emails to be answered. **Even as a farmer, there's computer work to be done** .... It's not uncommon to have meetings with other organizations around the city. Only about half my day is taken by work with plants, the other is taken by work with people.

LM What do you like best about your job?

DS I usually joke that **the things I love best are raising bees and raising interns.**

LM What are the biggest challenges?

DS One of the things is that there's no how-to manual on this stuff. I'm always trying to solve the problems that come up — it's always a challenge. And **how do you make it work not just horticulturally, but also economically** — how do you make it a business?

LM How important is the relationship between the farm and the restaurant below?

DS I don't think there's anything that's more important. I really don't. From my perspective, **the closer the coordination between the chef and the farmer, the better things work altogether.** With a hundredth of an acre, I have to make sure that everything I pull off the roof the chef will use .... I have to make sure that the chef is excited about what I'm growing, and that it's a positive experience for the customer.

LM To what degree does the restaurant subsidize the farm?

DS Financially, the value of the farm more than pays for all the inputs to the farm, the tools we buy for the seasons, all the direct costs. It just doesn't pay for the labor.

LM How involved is the restaurant's owner?

DS Really involved. We have meetings every week, I run my crop plan past her several times [prior to planting]. She's very, very hands on.

LM What marketing initiatives have been deployed?

DS There's almost no advertising whatsoever .... [Almost] everything else is done through PR, partnerships with other business ... the website and **social media.**

> "Only about half my day is taken by work with plants, the other is taken by work with people."

> "I have to make sure that the chef is excited about what I'm growing."

Restaurant, Uncommon Ground, IL

▌ urban apiaries
Philadelphia, PA

the farm

Honey from Urban Apiaries is all the buzz in Philadelphia's local food scene. The company tends bees at seven apiary locations around the city and sells raw (unpasteurized) honey throughout the greater Philadelphia area at farmers' markets, specialty shops, food co-ops and one high-end garden center. Several of these apiaries are perched on rooftops, such as those at Milk and Honey Market in West Philadelphia and Weavers Way Co-op in the city's Chestnut Hill neighborhood.

Apiarist Trey Flemming and business partner Annie Baum-Stein cofounded the company in 2009. At present, Flemming runs the business independently, tends the company's hives and processes the honey. I first met him in 2011 on the roof of SHARE Food Program, a food distribution center and urban farm in North Philadelphia. At the time, SHARE housed 14 of the company's 32 hives. Flemming worked on the roof with an assistant, who helped to haul equipment, carry supers from the hive and perform other supplementary tasks. Urban Apiaries also hires assistants for honey processing, which occurs at Flemming's farm in rural Pennsylvania.

After years of beekeeping in the agrarian landscape, Flemming finds that his urban bees are actually healthier than their rural counterparts. In the countryside, bees often visit agricultural crops, which are commonly doused with chemical pesticides. By contrast, urban bees graze on local flora found within street plantings, parks, roof gardens and vegetable gardens. These small-scale polycultures are generally not treated with synthetic pesticides, and so the visiting pollinators avoid bringing chemicals back to the hive. "On average, we probably harvest about 120 pounds of honey from each [rooftop] hive per year. The first season, we only harvested 90 pounds per hive," explained Flemming. These yields are higher than those of most rural hives, which makes urban beekeeping that much sweeter.

Perhaps the quirkiest quality of urban bees is their propensity to collect sugar by almost any means necessary. Bees can sometimes be found sipping from soda cans or munching on icing. Trey's bees at SHARE Food Program found their way into a nearby **jellybean factory**, and have since produced green and red cells of honey within their hives.

the farmer

Although Flemming is well versed in beekeeping, he seems to learn something new every time he works with his gentle pollinators. The bees' behavior in particular seems to captivate him, as do factors

*Trey Flemming of Urban Apiaries
on the roof of Philadelphia's SHARE Food Program, PA*

contributing to high production quality. He examined each super on the roof of SHARE with care, evaluating whether the honey was ready to be harvested, while looking for anything unusual throughout each hive. Flemming handled the supers calmly and confidently, while wearing a short-sleeved shirt. He does not wear a typical beekeeping jacket and veil, as his European honeybees (*Apis mellifera*) are less aggressive than other bee species. Flemming uses a smoker to calm the bees before he enters a hive, which involves puffing smoke from smoldering fuel (e.g., pine needles, leaves, wood chips) into the hive's entrance. Smoke helps mask the warning pheromones that bees emit when an intruder is approaching, which causes the colony to remain calm as the beekeeper enters the hive.

In addition to managing Urban Apiaries, Flemming owns and runs Two Gander Farm & Apiary, in Berks County, PA. This rural farm and apiary yield produce and honey that he also sells at market. Before establishing the farm with his wife, he worked as the farm manager at an organic vegetable farm in Chester County, PA.

secret to success

Urban Apiaries developed an **innovative branding strategy** that helps the honey sell like hotcakes. The apiaries are located in various neighborhoods throughout the city, and each jar is **labeled with the zip code** from which the honey was harvested. Philadelphians notoriously display great pride for their neighborhood, and branding by zip code taps into this innate loyalty. Customers generally prefer honey from their own zip code, and I've overheard shoppers on more than one occasion touting that the honey from their neighborhood simply tastes better than the rest. In fact, Urban Apiaries' honey is so popular that, in 2012, Flemming couldn't keep up with the demand — all this without a stitch of advertising!

When tasting the first drop of Urban Apiaries' honey, you'll further understand the obsession. The honey's rich, **complex flavors** change throughout the season. Each jar is unique in flavor, yet consistently irresistible. I eat it by the spoonful.

Flemming's expert approach to beekeeping further contributes to the success of Urban Apiaries. **Agricultural expertise**, in fact, is a reoccurring theme that contributes to the success of the other rooftop farm case studies. Rooftop farms are more expensive to construct and operate than ground-level farms. Efficient operations and high yields are therefore paramount in getting the most bang for your buck up on the roof. While rooftop gardens leave abundant room for experimentation, commercial rooftop farms benefit from an experienced farmer running the initiative.

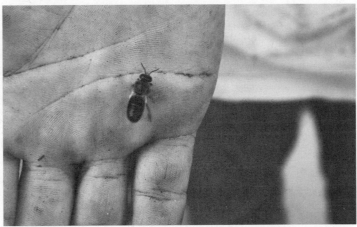

*Urban Apiaries operating on the roof
of Philadelphia's SHARE Food Program, PA*

Rooftop Farms [Medium-scale]

Trey Flemming interview

Urban Apiaries, Founder

LM How many hives does your company keep around Philadelphia?

TF We keep **32 hives in the city.** All together we have **seven apiaries,** and the honey is sold with [the apiary's] zip code on the label .... The largest apiary is here [in North Philadelphia on the roof of SHARE Food Program], with 14 hives. The smallest operation has two hives, but three hives is usually our minimum.

> **"The honey is sold with [the apiary's] zip code on the label."**

LM Have other building owners approached you to ask for bees on their roofs?

TF Yes, several building owners. [Everything from] restaurant owners [to] grommet shops and a local food co-op.

LM Is urban rooftop beekeeping a new practice?

TF No, bees have been raised on roofs in Paris for a long time.

LM What is the biggest challenge of keeping bees on roofs?

TF Public relations. Convincing people that the **bees aren't going to sting you.**

LM How much does a hive typically weigh?

TF About 300 to 350 pounds at peak production. That's if you don't harvest any of the honey .... On average, we probably harvest about **120 pounds of honey** from each [rooftop] hive per year. The first season we only harvested 90 pounds per hive.

LM Does honey production increase if you harvest honey from the hive?

> **"The more you harvest the more they produce."**

TF Actually it does. The more you harvest, the more they produce.

LM How often do you have to tend the hives?

TF Twice a week in the spring, because that's swarm season. I only come once every two to four weeks in the summer, which is much less often then I tend to my [rural] bees at home.

LM What is the tallest building on which you would keep bees?

TF The tallest building I've seen with bees was five stories, with bees essentially six stories up .... The building was surrounded by tall trees, which makes sense, since European honeybees are **canopy dwellers**. [European honeybees] are actually more comfortable on roofs than on the ground, because they can avoid predators [when on roofs] .... Some people have seen them all the way up in church steeples

LM Why do you keep your hives along the roof's parapet?

TF [The parapet] acts as a windbreak and protects the hives.

LM How well do urban bees overwinter, given that cities are fairly warm places?

TF They do pretty well. The [production season] is actually a bit longer in cities than it is in the country. Our bees did fine here [on the roof of North Philadelphia's SHARE Food Program] this past winter.

LM Would a rooftop farm benefit bees that already live on the roof?

TF It depends on what crops you plant. The **bees really love herbs**, especially anise hyssop, chives and also fennel. They seem to really like the okra here.

LM Would you ever have a rooftop farm in which you did not raise bees?

TF No. But I'm a beekeeper. I would face the hive entrances toward the [parapet] to divert the bees up and out. That way they won't get in anyone's way [who is working on the farm].

LM Given the expertise needed to tend hives, which do you think is a more viable model: rooftop farmers who tend their own hives, or beekeepers that travel and tend hives at all the rooftop farms?

TF Having a **network [of beekeepers]** is probably better because it takes a fair amount of knowledge to [tend hives] well.

> "It takes a fair amount of knowledge to [tend hives] well."

lufa farms

Montreal, QC

the farm

Perched atop an office building in Montreal's Saint-Laurent neighborhood, Lufa Farms feeds approximately 2,000 people every week. Since 2011, the 31,000-square-foot hydroponic greenhouse has provided over 25 varieties of fresh, delicious produce to a city that imports virtually all of its fruits and vegetables. The approach is simple. Customers buy a 12-week subscription for roof-fresh produce, which is delivered weekly to one of over 50 pick-up locations around the city. Lufa Farms offers several produce baskets, as they're called, and individuals, families and groups select the basket that's right for them at the start of the 12-week period. For the summer 2012 season, produce baskets ranged from $22 to $42 per week, and included 8 to 13 pounds of produce, depending on the basket. Local Quebec farms supplement the larger baskets with additional produce that's more difficult to grow hydroponically, such as root vegetable, squash and berries.

This innovative approach to food distribution differs from the Community Supported Agriculture (CSA) model, in that the year-round growing season at Lufa Farms allows customers to buy in at any time, rather than just before the start of the season. Customers also have the flexibility to postpone a week's produce basket (if they're away on vacation, for example), which is generally not afforded with CSA subscriptions.

As a matter of practice, Lufa Farms does not use any synthetic fertilizer, pesticides or herbicides. Plants are started from seed in soil-less flats, and then transplanted into more spacious hydroponic growing channels as they mature. All crops receive a carefully calibrated recirculating nutrient solution and are pollinated with honeybees that live in the greenhouse.

the founder + president

Mohamed Hage, a Lebanese-Canadian in his early thirties, holds an impressive track record of entrepreneurial accomplishments. Prior to Lufa Farms, Hage founded Cypra Media, one of the largest full-service email-marketing providers in Canada. In 2007, he began dreaming big about rooftop agriculture with friend and Lufa Farms cofounder Kurt D. Lynn. The men's dissatisfaction with the taste and quality of produce in Montreal's groceries spurred them to take action by establishing Lufa Farms, along with cofounders Yahya Badran and Howard Resh. With a team of experts by his side, Hage oversees all operations at Lufa Farms and focuses specifically on research, planning and build-out. Hage can finally enjoy the flavors of fresh produce again, just like those in the Lebanese village where he grew up.

secret to success

Lufa Farms fulfills a need. In a city that relies upon food imports from across the province and globe, Montrealers jump at the opportunity to buy local produce. Customers have always known that fresh fruits and vegetables are more nutritious than imported produce, and now they can taste the difference, too!

The farm's success is also attributable to the division of its well-educated 35 staff into five teams. Each team oversees a different facet of the company, and together, they make Lufa Farms run like clockwork. The Research Team monitors crop varieties for taste and nutrient levels, and develops technologies and techniques that support multi-crop greenhouse production. The Growing Team, led by Dutch horticulturist and agronomist "Tiny" Van Poppel, runs day-to-day operations and is staffed by college-educated growers. The Consumer Team manages ordering, packing, distribution and delivery, while the Community Team works to educate the community about urban agriculture and solicit ideas about what consumers want. Lastly, the founders, comprising Hage and three talented colleagues, manage the company's business and development operations. With such a dynamic staff, no wonder Lufa Farms is on the map!

Founders Kurt D. Lynn, Mohamed Hage and Yahya Badran (Howard Resh not shown); Lufa Farms, QC PHOTO BY AND COURTESY OF LUFA FARMS

Mohamed Hage interview

Lufa Farms, Founder + President

LM When did you start thinking seriously about developing your own rooftop farm?

MH I came to Canada when I was 12 years old from a village in Lebanon where everyone grew their own food .... Everyone said it's too cold to grow food in North America .... Then about six years ago, I said to myself, if you're going to address temperature, proximity to consumer, etc., you could revolutionize agriculture, and make it not just sustainable, but also profitable.

LM How many full-time staff do you employ at Lufa Farms?

"We only have two full-time growers, who grow the food for 2,000 people."

MH We have about 35 people in the company right now. The bulk of the team is in operations, development, [research and development], etc. .... We only have two full-time growers, who grow the food for 2,000 people. Two more people pack all the [weekly produce] boxes and distribute the food .... We are forming this [farm model] to be reproducible.

LM How are microclimates achieved within the greenhouse?

MH We are able to provide several microclimates within the greenhouse to provide the plants with exactly what they need .... Multiple irrigation systems feed the various crops and deliver specific nutrient levels to each.

LM What was the total project cost?

MH The construction cost was about $2.2 million .... This was a hard project to finance .... My family had to put up [almost] the entire investment .... [Eventually] we got one bank to give us a small loan, and a group that promotes economic development in Montreal also invested .... The SBEN contributed as well.

LM What is the farm's estimated payback period?

"Our costs are much lower than [a traditional ground-level farm]."

MH Whenever we build a new farm, we expect the payback to be 3 to 5 years .... Our costs are much lower than [a traditional ground-level] farm. We use about half the energy... and we spend [only] $15 a day in fuel to deliver food to over 2,000 people. We have virtually no distribution costs because all of our customers live within 5 to 10 kilometers [of the farm].

Growing Team employees tending to tomato plants (above)
atop a fully functioning office building (below); Lufa Farms, QC
PHOTOS BY AND COURTESY OF LUFA FARMS

Rooftop Farms [Medium-scale] 133

LM What are your current yields?

MH We currently achieve commercial-level greenhouse [hydroponic] yields. We can feed one person continuously with roughly 15 square feet.

LM What has been your greatest challenge at Lufa Farms?

MH People look at [Lufa Farms] as a farm on the roof, but it's a development of a new typology. Like any typology, there are challenges when translating a prototype into an actual functioning facility.

LM How have you made this transition easier?

MH We developed a software system that marries our production with our distribution …. Every tomato ever harvested was delivered the same day it was harvested …. Not only do we grow locally here at Lufa, but we're able to preserve the unbelievable quality of the produce [because we sell directly to consumers].

> "Every tomato ever harvested was delivered the same day it was harvested."

LM What has been your greatest success?

MH I'm a subscriber of Lufa Farms, and to me, the ultimate success is opening the [produce] box every week and knowing that [the food] was grown sustainably, by people that I know …. Now I know who my farmer is, and this is the start of something that could be very, very big. This is the biggest success of my personal life.

LM What is your favorite vegetable?

> "We hope that we're developing something that will not only feed people, but will revolutionize agriculture."

MH By far, arugula. I love it. Every sandwich must have it.

LM Does current demand for your produce necessitate a second location?

MH Absolutely. We're hoping to start construction [on a second rooftop greenhouse facility] before the end of the year in Montreal. It should be up and running in not too long. We hope to more than double our current capacity. We hope to use this concept all over North America, and all over the world …. We hope that we're developing something that will not only feed people, but will revolutionize agriculture.

Peppers from the 31,000 square foot greenhouse,
Lufa Farms, QC

farm checklist

Selecting the building on which to site your commercial rooftop farm can be daunting. Where should you look? What should you look for? What conditions will put the project over budget? How do you get the most bang for your buck? While your farm will provide many benefits (fresh food, green space, marketing potential, etc.), keep in mind that it is a business, and must therefore remain economically sustainable.

The Farm Checklist is designed to help you realize your rooftop vision by troubleshooting problems before they start. It will shepherd you through the process of selecting a prime location that contains the rooftop infrastructure your farm needs to succeed. The Farm Checklist is most effective when used during the initial stages of planning. This tool can reduce risk and save you money by minimizing roadblocks that commonly occur during design, construction and the inaugural months of your rooftop endeavor.

This roof was chosen over neighboring sites due to its structural integrity, Eagle Street Rooftop Farm, NY

☑ **Zoning + building codes**

  ☐ Agriculture permitted in relevant land use zone

  ☐ Local building codes reviewed and understood

☑ **Microclimate**

  ☐ Exposure considered when selecting building

☑ **The building below**

  ☐ Structural capacity approved by engineer

  ☐ Rooftop access meets building code requirements

  ☐ Parapet surrounds roof and is at least 48 inches above
     finished grade

  ☐ Waterproofing membrane approved by waterproofing provider

☑ **Up on the roof**

  ☐ Water hookup available and fully functioning (with enough psi)

  ☐ Space available for amenities

☑ **Dollars + cents**

  ☐ Necessary labor force is well understood

  ☐ Marketing plan clear and defined

  ☐ Profits and long-term financing understood

1. zoning + building codes

The critical first step in the Farm Checklist is understanding which neighborhoods in your city allow for agriculture. Local **zoning maps** delineate the specified "land use" for each area of the city (e.g., residential, commercial, light manufacturing, etc.). The local **zoning code** corresponds to this map and describes the acceptable uses for each land use type. Agriculture is only permitted under certain land use types. Be sure to read the fine print in the zoning code that is specific to your city, as **acceptable uses** vary by city. Some formerly restrictive cities, like Philadelphia, are currently revising their antiquated zoning codes. New codes will hopefully foster the development of rooftop agriculture across the city.

After selecting a neighborhood, take a close look at the local building codes. By considering snow load, roof access, parapet height and the use of reclaimed water for irrigation, these codes will help you select the right building for the job. The following pages discuss these specific building codes in detail.

*Winds speeds reached 50 mph during green roof construction,
Lower Manhattan hotel, NY*

2. microclimate

Step two is understanding the local microclimate. The importance of this became overwhelmingly apparent as I struggled to keep my footing atop a high-profile Manhattan hotel. This hotel sits one block from the Hudson River. From six stories up the view is unparalleled — and so is the wind. During construction oversight at the hotel, the green roofs and adjacent raised bed production area experienced **50 mph winds**. The street level contained a much milder microclimate, but up on the roof, temperature fluctuations and extreme wind conditions reigned.

Exposure is one of the most significant obstacles to rooftop farming. High winds cause winnowing (soil loss) and desiccation (soil drying), while temperature fluctuations can cause crops to bolt (flower) prematurely. Each roof experiences a slightly different microclimate, but some basic rules of exposure are as follows:

1) **Mind the water:** Rivers and other bodies of water often act as wind corridors in cities. Selecting a roof that is shielded from these channels can help to minimize extreme rooftop winds.

2) **Stay low:** Higher stories generally experience greater wind speeds. This means that a farm on top of a one- to three-story building will experience less stress than one on a taller structure.

3) **Surround yourself:** Take advantage of high neighboring buildings and taller segments of the farm's own building that can act as windbreaks. Positioning your farm directly south of a taller wall can help to block gusts. The wall may also capture heat, which will warm the adjacent soil. Be sure to also avoid taller buildings to the south (which could cast shadows on your farm), as well as vacant lots to the south (where a taller building could be erected).

One serious rooftop threat that ground-level farmers will never face is reflectivity. Light that reflects off surrounding surfaces can scorch spinach and burn brassicas. This means that building next to a reflective glass curtain wall should be avoided. Here are two anecdotes that illustrate the power of reflectivity:

"That's One Hot Tomato"
The green roof firm for which I work designed and built a courtyard shade garden for an important Philadelphia client. The courtyard is surrounded by taller building stories, which are faced with glass so that workers can enjoy the garden view. The garden's ferns, heuchera and other shade plants performed well at first, until strong summer rays reflected off the windows and fried some of the plants! What seemed at first to be a shaded haven had become a

seasonal hotbox. The most sensitive plants were replaced with sun-loving species, and next time we will take care to avoid a similar mishap.

"What Happens in Vegas ..."
The second anecdote involves a Las Vegas hotel that found itself in a hot spot when poolside guests were burned by light reflecting off the hotel's glass façade. For 90 minutes each day, the concave building reflects light that is hot enough to melt plastic and burn hair. Guests have the ability to seek refuge under patio umbrellas, but replace people with vegetable plants, and you'd have some fried green tomatoes on your hands.

4) **Cover up:** Temperature fluctuations can be minimized by covering your crop rows with shade cloth. This thin cloth is used regularly on ground-level farms, and it benefits crops by capturing the heat that is released by the plants and soil. Because rooftop wind will fill the cloth like a sail, it's best to build low hoop houses to frame the cloth. These hoop houses should be screwed or bolted to the sides of raised beds, or ballasted by the walkways between farm rows.

Reflective glass curtain wall on a neighboring building, Lower Manhattan, NY

3. the building below

structural capacity

Now that you've selected a site for your rooftop farm (with building codes and microclimate in mind), you must must *must* hire an engineer to assess the building's structural integrity. The weight of a rooftop farm can't be supported by most buildings, and so this step is extremely important — don't skimp!

The **structural engineer** may initially assess the building by examining any "as-built" architectural drawings that are available, but will ultimately need to conduct an onsite evaluation. The evaluation will be minimally invasive so long as the building's structural members (columns and beams) are exposed from the underside. If the members are not exposed, then the engineer may need to cut into the building to evaluate its construction and determine the load capacity.

The engineer should provide you with: 1) **dead load** capacity (how much your fully saturated farm can weigh per square foot); 2) live load capacity (how much weight is allotted for people and moveable items per square foot on the roof (OSHA requires 100 psf for people on accessible rooftops); 3) **snow load** (how much snow the roof can support; this is specified by local building codes); 4) **point loads** (localized areas that can support additional loads, like planters); and 5) **line loads** (linear locations that can support additional loads, like rows of raised beds).

If the building isn't strong enough to support a rooftop row farm, then consider using raised beds or containers instead. If the building can't support any additional load, then you can opt to structurally reinforce the roof (which will be very expensive) or find a new site. Just remember, the hassle of finding a new site is always preferable to collapsing a roof, particularly one that you may not own. The decision is a no-brainer.

rooftop access

Once the structural engineer gives you the go-ahead, it's time to think about access. How will people safely access the roof, and how will they transport materials up to and down from the farm?

Local building codes will likely dictate how you can access the roof. These codes may require an exterior, or even interior, **staircase** for regular rooftop access. If an interior staircase is required, then you might want to think seriously about selecting a building with a staircase already in place. Building an interior staircase that provides roof access can be quite expensive.

Some buildings contain an **elevator** that extends up to the roof. This piece of building infrastructure is ideal for larger agricultural operations, where traversing a staircase repeatedly can be extremely inefficient and tiring. Freight elevators are preferable to public elevators, since your materials and harvest may be a bit messy.

You must must *must* hire an engineer to assess the building's structural integrity.

Local building codes will likely dictate how you can access the roof.

I have yet to see a construction elevator permanently installed on the exterior of a rooftop farm building, but this option is worth investigating. A construction elevator would provide rooftop access for people and materials, thereby eliminating the need to trudge through the building.

parapet condition

A parapet is a wall-like structure that surrounds many flat roofs. Local building codes specify a parapet height for roofs that are regularly accessed, generally 48 inches above the roof's walking surface. This means that a 60-inch-high parapet will measure 48 inches, once a 12-inch-deep walking surface or planting area is installed. Parapets can be built on existing buildings that lack these structures, but at a high cost. It's best to select a building that already has a high parapet.

waterproofing membrane

Every flat roof is covered with a water-proofing layer (known as a membrane) that keeps water out of the building below. These membranes are made of several different materials, some of which are more advantageous than others for a rooftop farm application. An ideal membrane, such as PVC, is durable and has passed FLL Guideline testing for root-resistance. The FLL Guideline sets national material and construction standards for green roofs in Germany — the global leader in green roof technology.

An existing building will already have a waterproofing membrane in place. The condition of this membrane must be assessed by a green roof designer or waterproofing provider representative to see if repairs or a new membrane is necessary. Repairs may include patching, rewaterproofing damaged areas or adding tapered insulation to areas that pond water. Most building codes dictate that no more than two layers of waterproofing can exist on a roof. This means that if a building already has two "roofs," or waterproofing membranes, then both membranes must be removed before a third roof is installed. This process is known in the industry as a **tear-off and reroof.** Whether repairs or a reroof is necessary, it is much less expensive to address waterproofing flaws before the farm is installed. Take this step seriously.

If you need a new membrane, then ask the waterproofing provider about your warranty options. Ten-, 15- and 20-year waterproofing warranties are standard within the industry, and you may be able to pin down a **single-source warranty** between the waterproofing provider and the green roof provider. This type of warranty will save you headaches (and money) down the road, should a leak occur during the warranty period. The warranty should include "uncovering," which means that the waterproofing installer will cover all costs associated with removing the soil, plants and other layers to find leaks and repair the membrane, should a leak occur. If uncovering is excluded from the warranty terms, then the farm's owner is responsible for this cost, which can be quite hefty.

Irrigation and space for flowering perennials helps to turn this space into a place, Eagle Street Rooftop Farm, NY

4. up on the roof

water hookup

No matter how large or small your farm, you will need rooftop access to water. It is possible to retrofit a rooftop with a "point of connection," but most plumbing costs can be avoided by selecting a roof that already has a connection.

To the best of my knowledge, all of the commercial rooftop farms in the US are connected to a **municipal water** line. This connection provides a reliable water supply that can be tapped into as needed.

Connecting instead to a **large cistern** (in the building's basement or on the ground next to the building) presents an interesting option for water reuse. Using harvested rainwater for irrigation is a fabulously sustainable strategy, so long as: 1) building codes allow for roof runoff to be used to irrigate food crops (a filter may need to be installed on the cistern); 2) the cistern is large enough for anticipated water needs (remember that media will not hold as much water as soil, so more irrigation will be necessary than with ground-level farms); and 3) the cistern pump is powerful enough for the water to reach the roof (irrigation systems require a minimum amount of pressure by code).

In addition to irrigation, you may want a second point of connection for a wash station (for harvested produce). It's best to flesh out your design and think through all of the details prior to construction.

A **landscape architect** or **green roof designer** is typically hired for this stage of development. She will save you time and angst, and her accumulated experience will enable her to propose concepts that you have not yet considered. She will also coordinate with other trades (such as the architect, structural engineer and MEP engineer) to ensure that your system is fluid and meets all code requirements.

space for amenities

Before meeting with a landscape architect or green roof designer, think about all of the non-farming items that are essential to your farm. These "amenities" should reflect all anticipated user needs, and will turn your farm from a space into a place.

Functional amenities may include a composting area, tool shed, rooftop apiary and small greenhouse (for starting seeds). **Additional amenities** may include a shade structure and gathering space or demonstration area. You will also likely need a wash station. Large amounts of produce (for markets, restaurants, CSAs, etc.) should be processed at a facility that is inside the building or off-site, as rooftop real estate is too valuable for a large washroom. A small rooftop wash station can

[Rooftop] amenities ... will turn your farm from a space into a place.

be useful during demonstrations so that visitors can harvest and sample produce without having to leave the roof. A garden hose can always be used as a crude wash station if space or budget is constrained.

Brooklyn Grange Head Farmer Ben Flanner working a farmers' market stand, Brooklyn, NY
PHOTO BY JAKE STEIN GREENBERG
COURTESY OF BROOKLYN GRANGE

5. dollars + cents

labor force

No farm can succeed without a knowledgeable, hardworking **farmer**. In the face of shifting market values and unpredictable weather, a farm's pulse depends upon the farmer's unwavering dedication. This is a hard job. It is a stressful job.

But this is true of farmers at any altitude. What makes rooftop farmers unique? In today's boutique industry, high-altitude farmers are responsible for more than just growing food. Often times they must run workshops, organize volunteers, answer to the media and attend promotional events, all on top of their baseline agrarian duties. In this hip industry, a few rooftop farmers have the added responsibility of acting as role models — perhaps even icons.

Regardless of the farmer's celebrity, this hardworking individual is often seen as the **face of the farm**. Knowledgeable visitors tend to associate each rooftop farm with its primary farmer, even though this person is likely neither the property owner nor the farm's founder. Some farmers are shareholders in the operation, but as rooftop farms continue to sprout up, farmers will increasingly be hired as employees, without the potential for partial ownership. Most of today's rooftop farmers are paid either a salary or a stipend. In the case of boutique rooftop farms, the farmer often serve as the entity's only paid employee.

With only one or two paid farmers, how does all the work get done? Why, with a **volunteer labor force**, of course! Since the North American rooftop farming industry is still in its infancy, these rare landscapes often act as magnets for volunteers who are curious, believe in the cause or simply want to get their hands dirty.

Most volunteers offer their time on a single occasion. Perhaps they come as part of an organized group (like on a school trip), or they just want to see firsthand what all the fuss is about. These individuals are essential to the functioning of a boutique rooftop farm, but none is so valuable as the **repeat volunteer**. This is a person who volunteers his or her time on multiple occasions, or perhaps even on a regular schedule. This type of volunteer will show up with a basic understanding of how tasks must be performed, and he or she will generally require less instruction and supervision than a first-time volunteer.

It is important to remember that this free volunteer labor force comes at a slight cost. Training, supervision and scheduling all require resources. The farmer may be too busy to take on these responsibilities, which means that a **volunteer coordinator** is often a valuable asset to any volunteer-oriented farm. When starting a rooftop farm that incorporates volunteer labor, be sure to consult with a labor lawyer and your insurance broker in order to fully understand your assumed liability.

High-altitude farmers are responsible for more than just growing food.

This free, volunteer labor force comes at a slight cost.

Brand label for value-added products,
Eagle Street Rooftop Farm, NY

marketing

Small organic farmers across Canada and the US are realizing the power of marketing. When consumers can clearly see where their farm-fresh food comes from, they begin seeking out specific growers. Similarly, some restaurants with local organic fare are beginning to list the farms that supply the ingredients in their menus. Getting a farm's name out there is key in this competitive marketplace, regardless of whether the farm is on the ground or three stories up.

One way to **spread the word** is through sheer volume. The California-based Earthbound Farm Organic brand, for example, sells such high yields of produce that their label is recognizable on grocery shelves across North America. Singular rooftop farms cannot compete with this kind of volume. Consequently, rooftop farms must use innovative marketing strategies to earn name recognition and gain a loyal customer base.

Several rooftop farms, including Brooklyn Grange and Eagle Street Rooftop Farm, are delving into the world of **value-added products**. These are minimally processed goods such as jellies or hot sauces, which are made partially or entirely from rooftop crops. The principle is simple: rather than growing a bell pepper that will be sold, eaten and forgotten, you can grow crops like hot peppers that can be used to make products that will survive in someone's kitchen for several

months. When **branding** a hot sauce jar with your farm's name and **logo**, you will be exposing consumers to your brand for the life of the product. Each trip to the fridge brings a spoonful of marketing potential. Producing value-added products is one of many strategies that can be used to ensure that your limited product volume goes a long way in the marketplace.

Another way to spread the word about your rooftop farm is by **allowing visitors** to experience the space for themselves. In telling others about an unforgettable rooftop experience, farm visitors can be valuable in marketing a brand through word of mouth.

The **media** can also be instrumental in spreading a brand. Print and broadcast media coverage of a rooftop farm can reach a wide audience. Within certain demographics, social media may go even further. Regardless of whether a newspaper journalist, cinematographer or blogger visits your rooftop farm, it will behoove you to be accommodating and put your best foot forward. The media's relationship with a rooftop farmer, as with any public figure, can be either helpful or harmful in building the farm's reputation.

Over the past few years, Trey Flemming from Urban Apiaries has learned how **positive media relations** can be an effective means of brand dissemination. Flemming and his urban honeybees often appear in magazine and newspaper articles around Philadelphia, as well as

Each trip to the fridge brings a spoonful of marketing potential.

in online articles, blogs and documentaries. The media loves Flemming, and consequently, his brand has reached corporate America. While once available only at local co-ops and specialty stores, Flemming's honey is now available at Terrain, Urban Outfitters' garden center.

Annie Novak, from Eagle Street Rooftop Farm, similarly embraces media coverage. When I visited the farm in 2011, Novak welcomed me to the roof while in the middle of a photo shoot on the other side of a kale patch. With a warm smile and welcoming air, it was no wonder that Novak was simultaneously engaging a film crew, journalist and me. The farm's **media relations director** alleviates the pressure by responding to media inquiries, scheduling appointments and assisting Novak with other related activities.

profits

Profitability ... depends ... on the ability to capitalize upon the boutique quality of the industry.

Profitability for today's rooftop farms depends in part on the ability to capitalize upon the boutique quality of the industry. Rooftop farms are still scarce enough that their uniqueness alone attracts business. In other words, if a head of rooftop lettuce costs 50% more than one grown on the ground, a restaurant may still buy it in order to market a unique product on their menu. Selling rooftop produce to restaurants may make sense for many farmers. Most small to medium-sized **restaurants** require a variety of produce, which is typically what rooftop farms grow. Many restaurants are also willing to adjust their menus with the season, which suits the crop rotation schedules of many rooftop farmers. Additionally, the quantity of produce required by restaurants is often much lower than that of supermarkets. While most rooftop row farms and raised bed operations cannot produce high enough yields for supermarkets, they can easily feed restaurants.

Some commercial rooftop farms practice **community supported agriculture** (CSA) in order to ensure a steady income stream. This involves a group of shareholders who buy a portion of the farm's produce at the beginning of the season, and then receive a ration of produce each week. When rooftop farms cannot produce enough to support a full CSA, they sometimes partner with other farms to meet production demands.

Another form of profit may come from charging an **entrance fee**. At least one rooftop farm in New York City charges visitors and media personnel for each visit, which adds to the farm's revenue. Some community-oriented farms scoff at the idea of charging a fee, but for others, business is business. When demand is high and supply is low, there is money to be made. Farms practicing any ideology often charge for rooftop classes, **workshops**, seasonal events and other farm activities.

Heavily branded Brooklyn Grange honey jars (top),
Farmers' Market, NY
Australian film crew filming a documentary (bottom),
Eagle Street Rooftop Farm, NY

TOP PHOTO BY JAKE STEIN GREENBERG, COURTESY OF BROOKLYN GRANGE

checklist, check!

Now that you have a more thorough understanding of the considerations for starting a commercial rooftop farm, try filling out the **Farm Checklist**.

How do you rank? Are you able to check off every box? If not, then can you think of strategies for fulfilling the missing criteria? For example, if a waterproofing provider does not approve the building's waterproofing membrane, then is it in your budget to replace the membrane? Can you launch a capital campaign to raise money for this effort? Does it make more sense to select a different building altogether?

These are the kinds of questions that you should ask yourself (and your business partners, board or investors, etc.) early on in the process, in order to ensure that the planning, design, construction and long-term functioning of your project run as smoothly as possible. Every project contains unique advantages and obstacles, so use this checklist as a framework for discussing and troubleshooting the intricacies of your distinctive rooftop farm initiative.

☐ **Zoning + building codes**

    ☐ Agriculture permitted in relevant land use zone

    ☐ Local building codes reviewed and understood

☐ **Microclimate**

    ☐ Exposure considered when selecting building

☐ **The building below**

    ☐ Structural capacity approved by engineer

    ☐ Rooftop access meets building code requirements

    ☐ Parapet surrounds roof and is at least 48 inches above finished grade

    ☐ Waterproofing membrane approved by waterproofing provider

☐ **Up on the roof**

    ☐ Water hookup available and fully functioning (with enough psi)

    ☐ Space available for amenities

☐ **Dollars + cents**

    ☐ Necessary labor force is well understood

    ☐ Marketing plan clear and defined

    ☐ Profits and long-term financing understood

[Large-scale]

6 | Rooftop Agriculture Industry [Large-scale]

While still in its infancy, the burgeoning American rooftop agriculture industry is poised to establish the institutional framework necessary for promoting steady growth. This chapter considers rooftop endeavors that recognize the potential of replicable farm models and organized farm networks. One case study (I Grow My Own Veggies) exists as a single farm location, with the potential to spread its DIY model to other rooftops in the southeastern US. The other case studies (Gotham Greens, the Fairmont Waterfront, and Brooklyn Grange) are currently expanding their networks by branching to other roofs. With each success, the potential for growth within the rooftop agriculture industry expands. With each rooftop vegetable consumed and farm publicized, the industry's popularity soars.

Greenhouse attendant at Gotham Greens, NY
PHOTO BY ARI BURLING, COURTESY OF VIRAJ PURI

Flagship farm with the Manhattan skyline beyond, Gotham Greens, NY

PHOTO BY ARI BURLING, COURTESY OF VIRAJ PURI

gotham greens

Brooklyn, NY

the farm

Gotham Greens is revolutionizing the way we think about growing food in cities. In contrast to most rooftop farms, which emphasize community involvement or rooftop agriculture's boutique quality, this state-of-the-art hydroponic facility focuses on one thing: high-yield production. The 15,000-square-foot (0.3 acre) commercial greenhouse operation in Brooklyn's Greenpoint neighborhood leads New York City in rooftop yields and has inspired other budding companies to pursue rooftop hydroponics.

The facility consists of a steel-framed ridge and furrow greenhouse, bordered by two large photovoltaic (solar panel) arrays. The greenhouse itself contains three bays, which are connected internally to equalize growing conditions and maximize efficiency. Mechanized climate control and irrigation systems further increase efficiency, providing crops with a stable, enticing environment in which to grow. And grow they do. When I spoke with Gotham Green's cofounder and CEO in 2011, the company anticipated annual yields of 100 tons. At the time, the production focus was leafy greens — mostly lettuces, salad greens and herbs. These crops are fairly easy to manage, which increases a grower's ability to produce predictable quantities for wholesale customers. Gotham Greens sells to a variety of retailers and restaurants, including high-end grocers like Whole Foods Market and D'Agostino's. With sales on the rise, Gotham Greens is scheduled to open three new rooftop greenhouse operations in 2013, which will increase the company's acreage by 180,000 square feet (4.1 acres).

The expansion will benefit more than just the company's balance sheet. Retailers and restaurants gain access to larger quantities of highly marketable greens; consumers benefit from the increased availability of fresh local produce; and the environment wins as well! Gotham Greens produces 20 to 30 times the yields of typical ground-level field production, while using 20 times less water.

the CEO

Viraj Puri cofounded Gotham Greens in 2008 with business partner Eric Haley. The savvy businessmen enlisted Greenhouse Director Jennifer Nelkin in 2009 to manage greenhouse operations. As the company's CEO, Puri plays a key role in business development, raising capital and increasing brand awareness. He has developed and managed green building and renewable energy start-up companies in India, Malawi and the US, and also worked at a New York-based environmental engineering firm. With

Left to right: Co-Founder Eric Haley, Co-Founder and CEO Viraj Puri, Greenhouse Director Jennifer Nelkin, Gotham Greens, NY

PHOTO BY ARI BURLING, COURTESY OF VIRAJ PURI

a diverse international background, Puri adds a tremendous amount of energy and expertise to Gotham Green's collaborative leadership team.

secret to success

In 2011, I asked Puri why customers demand Gotham Green's produce, and he responded that people are drawn to the consistent **high quality**. The indoor greens are supple and delicious, and the facility's year-round production allows for **dependable** yields, even when other farms are closed for the season. Puri added that customers also enjoy buying local produce "not just for the sake of being local, but what local stands for — lower transportation impacts, higher quality, freshness and keeping dollars closer to home." [1]

But quality reliable products are required for the success of any sales company. What sets Gotham Greens apart from the rest? A phenomenal **business plan**! Puri and his partners treat Gotham Greens as a legitimate for-profit business, from planning, to day-to-day operations, to marketing and PR. While other farms rely upon grants and parent organizations for funding, Gotham Greens is powered by product sales and private equity. The company acquired enough upfront capital to build a state of the art facility, and through strategic planning and management, it now supports over 25 full-time employees in the Greenpoint location alone.

Viraj Puri interview
Gotham Greens, Cofounder and CEO

LM Why did you decide to use hydroponic growing methods?

VP We thought [hydroponics] lent itself well to an urban area primarily because it's higher yielding [than other agricultural methods]. We thought that if we were going to operate a commercial farm, we would need the higher yields …. It is also a **more space-efficient** form of agriculture. There is also the ability to do year-round production in a greenhouse.

LM What company designed the greenhouse?

VP We designed the greenhouse facility. Gotham Greens designed, financed, built and operates the greenhouse all in-house, independently.

LM How are the greenhouse footings attached to the roof deck?

VP **Bolted and flashed in** …. Structurally the greenhouse is designed to meet international and New York City building code requirements.

LM What was your biggest design challenge?

VP Everything! Just doing it on a roof. Getting all the permits … making sure that structurally everything was okay. [The biggest construction challenges were] permitting, working on a rooftop and craning all of the materials up.

LM What crops do you currently grow?

VP Leafy greens. Varieties of salad greens, lettuces and herbs …. The greenhouse is capable of growing a wider diversity of leaf crops, including things like mustard greens, kale, arugula [and even] strawberries, tomatoes, cucumbers and peppers.

LM How high are your yields, typically?

VP We are expecting over 100 tons [of produce] per year.

LM Have you experienced high demand for your produce?

VP We have, yes …. We sell to both retailers and restaurants, including Whole Foods Market and D'Agostino's.

> "[Hydroponics] lent itself well to an urban area primarily because it's higher yielding."

> "We are expecting over 100 tons [of produce] per year."

LM Why do customers demand your products?

VP High quality. Reliable **consistent year-round supply**. And [the fact that it's local] is also a factor … not just for the sake of being local, but what local stands for — lower transportation impacts, higher quality, freshness and keeping dollars closer to home.

LM How is the company funded?

VP We are **privately funded**, including both debt and private equity …. It is a for-profit company.

LM Can you achieve a competitive scale through replicability?

VP Each roof is different with its own unique design and construction challenges. It's certainly going to be easier going forward, having completed one greenhouse; however, I don't believe we can create a simple template to follow. It's the same challenge for any retrofit project.

LM Are you planning to expand your current operation?

VP We plan to **expand to other rooftops** in Brooklyn.

LM Is widespread rooftop agriculture viable in New York City?

VP Maybe …. What we're trying to figure out is whether it's **commercially viable**. It remains to be seen because we are one of the first groups doing this. There aren't a lot of precedents …. There are many ways to farm responsibly and sustainably, and I think that urban agriculture has a potentially huge role to play …. In terms of commercial farming viability, it remains to be seen …. In the early stages of anything, you have to assume a little more risk to see if something works …. This is a working prototype.

LM Does the success of rooftop agriculture require political, architectural and cultural changes?

VP Yes. I think that it's already happening. The Department of City Planning has held a number of hearings and workshops … to see how barriers can be reduced …. It's **moving in the right direction** …. It's going to take time to establish best practices …. The city is trying to extend green roof credits to rooftop agriculture too.

15,000-square-foot hydroponic greenhouse,
Gotham Greens, NY

PHOTO BY ARI BURLING, COURTESY OF VIRAJ PURI

i grow my own veggies
Sarasota, FL

the farm

I Grow My Own Veggies occupies the 3,000-square-foot (0.07 acre) roof of a glass factory in Sarasota, FL. Founder and farmer Vincent Dessberg owns the building and began growing vegetables and herbs on its roof in 2009. "The ultimate goal is to show people how easy it is to grow your own food," he explained, when I visited his farm in 2012.[2] Farming and gardening in Central Florida's sandy, nutrient-poor soil can be difficult, and Dessberg's open-air hydroponic facility offers a practical solution for the locale. Open farm hours and street-level visibility provide visitors and pedestrians alike with a view into how feasible rooftop farming can be in this region.

A chain-link fence surrounds the farm along the parapet, one story above street level. Vinyl-coated mesh (the same material as the shade cloth used to border tennis courts) drapes the fence and helps shield crops from both the wind and Florida's unrelenting sun. The farm utilizes a vertical, soil-less growing system consisting of Styrofoam containers, stacked four- to six-high up a support pole. A PVC feeder line connects to the top of the farm's 188 poles/container stacks and supplies an irrigation drip emitter line to the highest container in each stack. Holes in the containers allow water to drip down through one container into the next — an efficient irrigation strategy. According to Dessberg, "There are 6,000 to 7,000 plants going at any one time" on the farm. An impressive feat for such a small rooftop.

During I Grow My Own Veggies' fledgling months, Dessberg intended to attract 50 customers, ostensibly CSA members, to buy into the farm. The idea was to take crop requests, grow the produce and allow customers to harvest their own produce at their convenience. Harvesting consumes approximately 20% of Dessberg's labor expenses, and so allowing customers to "pick their own" would increase the farm's economic sustainability. Dessberg could not drum up enough interest, and so he now performs most of the production work by himself, with occasional help from volunteers. The farm is currently in a state of transition, whereby Dessberg's friend and general farm manager Don Gamin will assume the leadership role.

the farmer

Vincent Dessberg is a self-described serial entrepreneur. The 52-year-old Florida native capitalizes upon a few real estate investments and spends the rest of his time running the farm, dreaming up inventions and tinkering with prototypes. His latest creation is a recumbent tricycle, which he let me ride during my visit.

Rooftop farming is clearly not Dessberg's life's mission, nor is it his singular passion. He's a practical guy, and when he perceives the need for local food, a product or anything else, he snaps to. Despite his leaping from one project to the next, growing food seems to evoke fond childhood memories within Dessberg: "I've always kind of been curious about farming. My mom had a small garden plot when we lived in Massachusetts .... There's something to it that grounds you."

secret to success

Dessberg's farm is unique to Florida and, more than likely, to the whole southeast region of the US. The farm's ability to **demonstrate** that rooftop agriculture is feasible in this part of the country is monumental and will hopefully inspire other building owners and entrepreneurs to follow suit.

I Grow My Own Veggies is further successful in that its growing system is lightweight and can be easily retrofitted onto existing roofs. The system is easily **replicable**, with the potential to spread to many rooftops. Efficient use of space and water increases the system's success, particularly when built in the urban environment.

Founder and farmer Vincent Dessberg, I Grow My Own Veggies, FL

Open-air hydroponic strawberries,
I Grow My Own Veggies, FL

Vincent Dessberg interview
I Grow My Own Veggies, Founder

LM Why did you decide to use a hydroponic growing method?

VD I had to try to figure out the **lightest [growing method]**, because I didn't want my roof to collapse. So I went with Verti-Grow, which [consists of stacked] Styrofoam containers. They're filled with half perlite … and a little bit of coconut coir …. Most of the weight's in the water in the coconut coir …. There's a couple of thousand pounds of plants [up here], but the system itself doesn't weigh very much.

LM How common is this method in Florida?

VD There's quite a few **Verti-Grow** systems here [in Central Florida], mostly because the soil is so terrible. If you want to grow anything in the ground and do it organically … it takes four or five years to build [the soil] …. The soil here is all sand, and you're lucky if you have 2% or 3% organic [matter] to begin with.

LM What crops are you currently growing?

VD Strawberries, chard, parsley, basil, a couple of different leaf lettuces … watercress, onions …. I love to grow all sorts of weird stuff — I'll try anything I've never heard of …. Recently I discovered that I really love mizuna and nasturtiums.

LM Why is urban agriculture meaningful to you?

VD Because to me it's always made sense that if you're going to grow food you should probably grow it where people are, and not where they aren't.

LM Why did you decide to grow on the roof rather than the ground?

VD [The site] was downtown, it was **convenient**, it just made sense.

LM Why do you think sales outlets and customers demand rooftop produce?

VD **You can't get any fresher.** It's gotta [come from] within feet if not miles of where you're eating it. Most produce loses a lot of nutrients over time … and so rooftop produce is [more nutritious].

LM Why did you choose to name your farm I Grow My Own Veggies?

> "You should probably grow [food] where people are, and not where they aren't."

VD The ultimate goal is to show people how easy it is to grow your own food, and that **you can grow your own food.**

LM What do you gain by allowing the public to access your farm?

VD Out of your labor expenses, 20% is harvesting. So my theory was that since [the farm] was downtown, if I could just get 50 people out there who I could give keys to, they could come harvest whenever they wanted .... They would just let me know what they wanted planted ... and 50 people would have **access to a farm 24-7-365.**

LM How long is the farm's expected payback period?

VD **I was hoping for a normal return** — you know, five or seven years. [The farm] was more of an experiment, to see if it was doable, to see what the reaction was, to see what the numbers were. I've searched the Internet forever, trying to find actual numbers on hydroponic farming, and there's nothing out there .... I did years of research.

LM What would it take to be profitable at your farm's current scale?

VD Crop rotation planning is something that you have to be on top of .... One operator can grow 14,000 or 15,000 plants hydroponically [with proper rotation planning].

LM Can you achieve the scale that you need to be profitable through replicability?

VD Yeah I could probably do that .... One guy could easily manage three or four [smaller rooftop farms] if they were in a block or two from one another. He might be able to handle more if he really wanted to work at it.

LM Is widespread rooftop agriculture viable in Florida?

VD Very much so .... **If it's a flat roof, it should have agriculture on it**; if it's a slanted roof, it should have solar on it. There should not be an empty roof up there.

LM Are local policies helpful or harmful to your farm's success?

VD Government [and zoning] are the biggest [obstacles] to urban farming, period .... Even after the government writes into its policies that it's supposed to promote [urban] farming ... the left hand doesn't know what the right hand is doing!

Farm manager Don Gamin up on the roof,
I Grow My Own Veggies, FL

the fairmont waterfront

Vancouver, BC

the garden

Executive Chef Dana Hauser knows fresh food and is committed to growing it as close to her kitchen as possible. The Fairmont Waterfront hotel, where Hauser works, has a long-standing rooftop-kitchen partnership, dating back to 1995 when former Executive Chef Daryl Ngata founded the hotel's rooftop garden with herbalist Elaine Stevens. As a matter of fact, growing food on location and partnering with local growers is nothing new for the international Fairmont Hotels & Resorts company. Select locations began cultivating fresh vegetables and herbs over 20 years ago, and today, edible rooftop gardens exist at Fairmont Hotels & Resorts in Canada, the US and Singapore.

When speaking with me in 2012, Hauser explained that the rooftop transitioned primarily into a non-edible perennial garden (with flowers and trees) after Ngata left the Fairmont Waterfront. Upon her appointment as executive chef, Hauser made revitalizing the rooftop garden one of her top priorities: "This is something that was really important for us, to get the [vegetable] garden back up and running for our signature restaurants."3 Hauser's commitment to reintroducing vegetables, herbs and edible flowers to the hotel's rooftop provides a fresh spin to the menu and brings some well-deserved

attention to the kitchen. The 2,100-square-foot (0.05 acre) vegetable and herb garden exhibits a formal layout, with planting beds surrounded by neatly trimmed hedgerows. A second rooftop production area, called the Terrace Garden, is home to the hotel's honeybees, which produce 600 pounds of honey each year. In total, the rooftop provides diverse crops including tomatoes, carrots, gooseberries and figs. Hauser even grows rhubarb — a perennial crop that no other rooftop farmer I've met grows. Daily rooftop harvests provide 10% to 15% of the kitchen's ingredients, according to Hauser's estimates, thereby freshening the fare for the hotel's two restaurants, in-room dining and tea service.

the executive chef

Hauser has the distinction of being the first **female executive chef** in the Fairmont Hotel & Resort's 120-year history. Her unbridled talent, whisked with a passion for fresh seasonal foods, distinguishes her as a leader within the company who pushes the limits of local.

Raised in Newfoundland, on the east coast of Canada, Hauser began her schooling at the Memorial University of Newfoundland, where she studied psychology. Craving a career that merged food, travel and teamwork, she enrolled in the Candore College's Culinary Arts program, which led

Executive Chef Dana Hauser,
The Fairmont Waterfront, BC

to a seat in the Southern Alberta Institute of Technology's Professional Cooking program. Hauser worked in several Fairmont Hotel & Resort kitchens before landing her current role at the Fairmont Waterfront.

secret to success

Chefs at the Fairmont Waterfront plant and harvest their own ingredients. The seasonal vision for the menu dictates crop planning, and daily trips to the roof ensure that each chef is in touch with what's ready for harvest. This seamless integration of roof and kitchen produces inspired chefs who see firsthand where their food comes from and are committed to creating seasonal culinary creations.

From a broader perspective, the company's ability to replicate its rooftop garden model to properties around the world provides unparalleled opportunity for success. Fairmont Hotels & Resorts already owns countless buildings. All it takes to build a rooftop garden on an existing location is converting rooftop flower gardens into attractive production areas or building some raised beds. It's an easy sell, and more importantly, the cost of buying or renting a building is nonexistent. Perhaps other companies and corporations will similarly see the benefit in planting rooftop gardens across their properties and providing fresh food to employees, clients and customers.

The hotel's 2,100-square-foot roof garden, The Fairmont Waterfront, BC
PHOTO BY AND COURTESY OF
FAIRMONT HOTEL & RESORTS

Dana Hauser interview
The Fairmont Waterfront, Executive Chef

LM What crops are you currently growing on the hotel's roof?

DH Several varieties of heirloom tomatoes, turnips, carrots, red and white potatoes, leeks, edible flowers, several varieties of lettuces … arugula, sorrel, rhubarb, an assortment of herbs — especially rosemary, rosemary is my signature herb right now — basil, sage, thyme, oregano and five or six varieties of basil. [We also grow] bay leaf trees, three figs are coming in next week, an apple tree, grapes, raspberries and gooseberries.

LM Is there an apiary on the roof that provides the kitchen with honey?

DH Yes … our Executive Sous Chef takes care of the bees and is a third generation bee-keeper …. We use the honey in our pastry shop, and it's on our tables for tea service …. We have six hives on property that provide 600 pounds [of honey], which gets us through half the year. Now we have an extra 20 hives at the Honeybee Centre near the [Toronto] Airport, which provide an extra 4,000 pounds per year.

LM How do you coordinate between menu planning and crop planning?

DH I'm up there every single day …. The chefs in the [hotel's two] restaurants go up every day [as well] to pick edible flowers for garnishing the plates or to cut herbs …. We also give bee tours constantly. As much as we can, we like to use food from the garden …. The vision of the menu dictates what we plant in the garden. [Once the garden is more established], we will build a menu around … what's working. We also do daily specials and VIP events [using the rooftop ingredients].

LM Do you coordinate, consult or meet with the executive chefs from the other Fairmont Hotel & Resort locations with rooftop gardens?

DH Just the other day, [chefs from Fairmont] Pacific Rim were over. They came to see what we're doing so that they can expand [their own rooftop garden].

LM What rooftop-inspired dish that your restaurants prepared stands out to you?

DH We do an heirloom tomato salad with fresh basil, buffalo mozzarella cheese, and 100-kilometer croutons. The tomatoes and basil are from our roof … the croutons are from a local bakery that we source.

"The vision of the menu dictates what we plant in the garden."

"We do an heirloom tomato salad with fresh basil, buffalo mozzarella cheese and 100-kilometer croutons."

brooklyn grange
Queens + Brooklyn, NY

the farm

The blustery wind atop the Standard Motor Products building in Queens, NY didn't deter five young business partners from investing in a vision. Brooklyn Grange, a New York-based for-profit rooftop farming operation, opened its first roof to vegetables in 2010 and a second in 2012. The flagship location, a 43,000-square-foot (1.0 acre) row farm in Queens, proved so successful that a second 65,000-square-foot (1.5 acre) farm — the largest rooftop row farm in the US — sprouted up quickly for the team. Cofounders Ben Flanner, Gwen Schantz, Chase Emmons, Michael Meier and Anastasia Cole Plakias work together to run everything from operations and marketing to the company's value-added product line and consulting services.

"[Our goals are] to grow delicious healthy vegetables, spread enthusiasm for healthy flavorful vegetables, improve people's diets, educate children, green New York City, grow farmers and employ farmers!" explained Flanner, the company's head farmer and president, when I interviewed him in 2012.[4] And grow healthy vegetables, they do. The two farm locations raise an incredible diversity of salad greens, nightshade, hardy greens, herbs, legumes and root varieties, as well as over 40 heirloom tomato varieties! Both row farms are built atop a plastic root barrier (to prevent vegetable roots from penetrating the roof's waterproofing membrane), a thick protection fabric, a 25-millimeter reservoir sheet (for drainage), separation fabric and 10 to 15 inches of green roof media. The substrate contains a much higher organic content than does typical green roof media, and the farmers continuously add compost and other organics to further enrich the mineral soil. Brooklyn Grange's flagship farm kicked off its inaugural season with 720 cubic yards of media, which took six days to crane up and transport across the roof using buggies. The five cofounders spread the media themselves, with the help of friends and volunteers. In fact, these energetic partners perform almost all of the farm labor. Brooklyn Grange brings in several apprentices and interns each growing season, some of whom receive college credit in exchange for their efforts. Volunteers often occupy the roof as well, as do visitors and aspiring rooftop farmers.

the farmer + president

Ben Flanner, a slender man in his early thirties with a dark beard and signature leather fanny pack, fell into urban agriculture after a career in financial analytics and marketing. The Milwaukee native's love affair with food production began in New York, where his green aspirations have reached new heights.

The flagship farm overlooking Long Island City,
Brooklyn Grange flagship farm, NY

Photo by Allen Ying, courtesy of Brooklyn Grange

Flanner studied industrial engineering at the University of Wisconsin, a career that helps him think through agriculture in a systematized way. His first taste of rooftop farming brought him to Eagle Street Rooftop Farm in Brooklyn, NY (see Chapter 5), where he cofounded and managed New York City's first rooftop row farm with skyline farmer Annie Novak. Flanner learned a tremendous amount from this first farm, applying the lessons learned to both Brooklyn Grange locations. By tweaking design and operational factors, he and his team were able to increase the square-foot revenue from $2.46 (seen at Eagle Street Rooftop Farm in 2009) to $3.15 (Brooklyn Grange's 2010 projections), with the potential to reach $3.59 per square foot.[5] The largest "tweak-worthy" aspects of farming at Brooklyn Grange seem to deal with media and wind. Farming seven stories up causes Flanner and his crew to get creative with protecting crops against heavy gusts. A common tactic they use involves installing thousands of bamboo poles as windbrakes. At times the poles double as supports for climbers (like peas) and top-heavy plants (like tomatoes). Flanner troubleshoots on a daily basis, and the farms' unique conditions keep him on his toes.

Like other prominent rooftop farmers, Flanner exhibits a breadth of talents in addition to his knack for growing food,

Head Farmer + President Ben Flanner tending to seedlings in the rooftop hoop house, Brooklyn Grange flagship farm, NY
PHOTO BY ANASTASIA COLE PLAKIAS, COURTESY OF BROOKLYN GRANGE

saying, "I ... love farming for the many moving parts, the many problems that have be solved daily and the many hats that we have to wear to run both a business and a farm." The success of Brooklyn Grange attests to Flanner's aptitude for business, marketing and finance and his clear understanding of how the many cogs of his business must work together to achieve profitability. His ability to turn "beans to bucks" makes him one of today's most inspirational rooftop farmers.

secret to success

Brooklyn Grange's dynamic business model focuses on the triple bottom line, which values a balance between human, environmental and economic sustainability (see Industry Checklist: Financing, later in this chapter). To appease the human side of the business's sustainability, Flanner and his team aim to create farming jobs that offer living wages within the city. In terms of the environment, Brooklyn Grange deploys organic growing methods to produce vegetables and herbs without the use of synthetic fertilizers, herbicides or pesticides. While the farms are not certified organic, the growers respect the organic principles. And then there's economic sustainability. Like any successful business, Flanner and his team developed a strong business plan that utilizes diverse revenue streams and a robust branding campaign.

Brooklyn Grange sells to restaurants (all within a five-mile radius) and directly to consumers. The company's 25-member CSA provides a predictable revenue each year, as members pay for a season's worth of vegetables upfront. Direct consumer sales at street-level market stands are a bit more unpredictable, but serve an important role in increasing the brand presence. Brooklyn Grange further disseminates its brand by manufacturing and selling a wide range of value-added products through its sister company, BG Value Added Goods. The honey, beeswax candles, salsa, pickles, jams and signature hot sauce are all slapped with labels that prominently read "Brooklyn Grange." Although partially catering to niche markets, these products are universal enough to fetch a price at any market or specialty shop.

Additional forms of revenue include special events, such as Brooklyn Grange's annual Honey Fest, which caps off New York City's Honey Week and celebrates the legalization of beekeeping in the city. Farm tours and rooftop farm consulting diversify the company's revenue streams even further and come in handy year-round. The team feels confident enough in the business that they are in it for the long haul. Brooklyn Grange signed a 10-year lease with full rooftop access for its flagship farm and a whopping 20-year lease with the Brooklyn Navy Yard for the second location. This is one company that may just prove that rooftop row farming can be commercially viable in North America.

Flanner's ability to turn "beans to bucks" makes him one of today's most inspiring rooftop farmers.

Ben Flanner interview

Brooklyn Grange, Head Farmer + President

LM How did Brooklyn Grange's five cofounders begin collaborating?

BF The Grange team met in the summer of 2009. I was managing the Eagle Street Rooftop Farm [in Brooklyn, NY], and the team meshed at Roberta's restaurant in Bushwick. At the end of the season, we began putting together the business plan based on the sales data from the 2009 season [at Eagle Street Rooftop Farm].

LM Did the company initially intend on establishing multiple farm locations?

BF Yes, the vision has been to expand from the start.

LM How was each roof selected?

BF We select roofs based on location, structural integrity, accessibility and, of course, our ability to strike a deal with the [building] owners.

LM What crops have you had the most success with?

BF Salad greens — we take great pride in the quality of our greens mixes — they change through the seasons .... [We continuously test] new types of leaves to blend. Heirloom tomatoes are also delicious, plentiful and one of our more profitable crops.

LM What is Brooklyn Grange's most profitable crop?

BF Greens mixes.

LM How long is your anticipated payback period?

"We plan to pay back our initial investment within ten years."

BF We plan to payback our initial investments within ten years. We operate similarly to a manufacturing system, with an upfront investment, long-term business plan and slow and steady payback.

LM What are the primary goals of Brooklyn Grange?

BF To grow delicious healthy vegetables, spread enthusiasm for healthy flavorful vegetables, improve people's diets, educate children, green New York City, grow farmers and employ farmers!

Flagship farm with its signature water tower (top), and Co-Founder Gwen Schantz harvesting (bottom); Brooklyn Grange flagship farm, NY

Photos by Allen Ying (top) and Anastasia Cole Plakias (bottom), courtesy of Brooklyn Grange

LM What attracts you to urban agriculture?

BF I am attracted to beautiful vegetables. I also love farming for the many moving parts, the many problems that have be solved daily and the many hats that we have to wear to run both a business and a farm.

LM How is your experience with industrial design and financial analytics relevant to farming?

BF They are both extremely helpful. Industrial engineering is the practice of time and motion, design processes and accomplishing things efficiently. My financial background has been very important for keeping a **tight financial ship** at the farm, and has been crucial in building our business plan, attracting investors and convincing people of our legitimacy.

LM What lessons from farming at Eagle Street Rooftop Farm did you apply to Brooklyn Grange?

BF We made several tweaks to the farm, but used the same general principle that I designed with a series of growing beds and thin walkways in between. More practically, the data that I collected of all of our sales [from Eagle Street Rooftop Farm] were crucial to developing the business case [for Brooklyn Grange]. Rather than unknown revenues and costs, we had a smaller model with hard data to use.

LM What has been your biggest success?

BF Starting two new farms, receiving the Community Business Award from the Mayor's Office and seeing all of the beautiful vegetables leave the door.

LM What would you change in establishing a third farm location?

BF We have to work with our new farm for a bit longer, figuring out any limitations, restrictions, and we will be wiser soon.

LM What advice would you give to aspiring rooftop farmers?

BF Work hard, good things don't come easily! And make sure to have fun and crack some jokes while you're working.

Healthy crop rows with the Long Island City beyond,
Brooklyn Grange flagship farm, NY

Photo by Anastasia Cole Plakias, courtesy of Brooklyn Grange

industry checklist

One location not enough for you? Before you dive into a multi-farm endeavor or try to jump start your own revolution, run through the Industry Checklist to make sure you know exactly where you're headed. Developing a strong business plan and pinning down your financing strategy before you get started is essential, as the start-up capital required for a single rooftop farm can easily reach $100,000, or even $2 million in the case of hydroponic greenhouse operations. The Industry Checklist can also support city planners, policy makers and developers, as they consider rooftop agricultural strategies on a neighborhood or citywide scale.

Farm-side market stand, Brooklyn Grange Brooklyn Navy Yard Farm, NY
PHOTO BY JAKE STEIN GREENBERG, COURTESY OF BROOKLYN GRANGE

☑ **Primary objective**

    ☐ Underlying purpose of initiative determined

    ☐ Long-term goals identified

    ☐ Milestones and measures of success mapped

☑ **Financing**

    ☐ Triple bottom line considered

    ☐ Business plan clearly defined

    ☐ Profits and payback period understood

    ☐ Funding opportunities reviewed and understood

☑ **Location, location, location**

    ☐ Climate limitations understood

    ☐ Local zoning and policies understood

    ☐ Proximity to "eaters" evaluated

    ☐ Building stock assessed

☑ **Decision Tree**

    ☐ Tree consulted when selecting production methods

☑ **Forging relationships**

    ☐ Universities and research institutions contacted

    ☐ Partnerships with distributors and food aid entities explored

    ☐ Community education and outreach planned

1. primary objective

mission

What is your company's primary goal? To address food justice? Fill a high-end niche market? Provide fresh produce to as many people as possible? Distilling the precise purpose of your company is key to developing clear and simple messaging. Your targeted consumer groups will not understand your company's purpose unless *you* understand it first. Here are a few examples of company objectives:

"To educate people about how to grow food in the urban environment, empower people to make healthy food choices through understanding our food system, and to demonstrate low-cost container gardening." — Lindsey Goldberg, Graze the Roof (see Chapter 4)

"[To realize] the benefits of green roofing while bringing hyper-local produce to the North Brooklyn community." — Eagle Street Rooftop Farm website (see Chapter 5)

"[To produce] the finest quality, freshest, best tasting, and most nutritious culinary ingredients available in New York City." — Gotham Greens website (see Chapter 6)

"To create a fiscally sustainable model for urban agriculture and to produce healthy, delicious vegetables for our local community while doing the ecosystem a few favors as well." — Brooklyn Grange website (see Chapter 6)

Some companies choose to reevaluate and adjust their mission every few years (such as at five-year intervals) to stay current within an evolving marketplace.

long-term goals

Establishing several long-term goals can help to break down your company's mission into bite-sized morsels. This involves outlining specifically how you will fulfill the mission, a task that will ultimately lay the groundwork for establishing business milestones and measures of success. Long-term goals can be as simple as: A) Provide fresh, delicious food to the local community, B) Raise crops using organic practices and C) Minimize the company's carbon footprint. These goals are fairly general but will plug in nicely to the next step.

milestones + measures of success

Next, try establishing quantifiable milestones to carry out each long-term goal. The most effective milestones are paired with dates, such as "by the by the end of the first year of business." Here are some examples of milestones that pertain to the goals listed above. Try drawing a similar diagram for your own rooftop farming company.

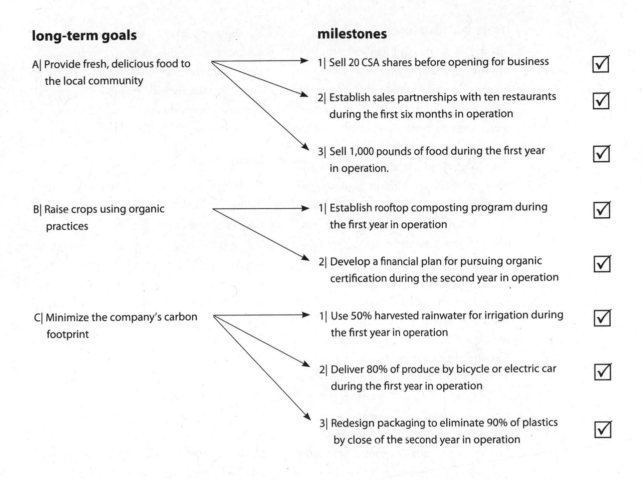

long-term goals

A| Provide fresh, delicious food to
the local community

B| Raise crops using organic
practices

C| Minimize the company's carbon
footprint

milestones

1| Sell 20 CSA shares before opening for business ☑

2| Establish sales partnerships with ten restaurants
during the first six months in operation ☑

3| Sell 1,000 pounds of food during the first year
in operation. ☑

1| Establish rooftop composting program during
the first year in operation ☑

2| Develop a financial plan for pursuing organic
certification during the second year in operation ☑

1| Use 50% harvested rainwater for irrigation during
the first year in operation ☑

2| Deliver 80% of produce by bicycle or electric car
during the first year in operation ☑

3| Redesign packaging to eliminate 90% of plastics
by close of the second year in operation ☑

*Diagram of potential long-term goals and
milestones for a rooftop farming company*

2. financing

triple bottom line

The major players in today's rooftop agriculture industry understand business. They understand how to get projects off the ground (pun intended) and position themselves successfully in today's competitive marketplace. How do they accomplish this? Well, most of these companies make business decisions with the triple bottom line in mind: people, planet and profit.

This business mantra acknowledges humans, the environment and profitability as three interlinked pillars of sustainable business practice. The philosophy asserts that a business cannot survive long-term without sustaining all three pillars to ensure social, environmental *and* economic sustainability. In practice, every business decision must consider the impact on each of the three pillars. For example, if the opportunity arises to increase a company's capital tenfold by laying off half the employees and dumping toxic pollutants into a river, profitability will be affected positively while the other pillars will be deleteriously affected. After considering people, planet and profit, the company will likely decide not to move forward with the initiative. Other decisions are not so clear-cut. Operating according to the triple bottom line can sometimes prove to be a balancing act, wherein compromises between the pillars are negotiated. Rarely does a decision affect all three pillars in the same way, to the same degree. When operating according to this, or any model, keep in mind that your business can only continue eliciting positive change and providing valuable services so long as it remains economically sustainable.

business plan

Crafting a business plan may seem intimidating, but thinking through all aspects of your endeavor before getting started will help you establish a game plan for success. Business plans articulate the business's organizational structure, operations, financing and often times marketing, while providing payback projections and revenue milestones. Successful business plans clearly describe the company's business model, which should remain flexible enough to adapt to changing external conditions over time. Periodic (often quarterly) checks are often built into the business plan as well, to regularly evaluate the company's performance relative to projections and milestones.

Not to fear if business isn't your *forte*! Most commercial rooftop farms with more than one location leave production to the farmers and bring at least one business wiz onboard to run financial operations, business development and strategic planning. Viraj Puri, a savvy CEO with a background managing start-up companies, runs Gotham Greens, a multi-farm hydroponic operation on New York City rooftops that produces greens and herbs for Whole Foods Market and other high-end markets and restaurants. BrightFarms — a company that designs, builds, finances and manages hydroponic greenhouses on top of and in association with supermarkets throughout the US — brought on CEO Paul Lightfoot to grow the business and develop strategic partnerships. Mohamed Hage and Kurt D. Lynn, two out of the four cofounders of Lufa Farms — a Montreal-based hydroponic greenhouse operation expanding to multiple locations throughout North America — come from business-related backgrounds as well. Even Brooklyn Grange's head farmer and president Ben Flanner entered into rooftop farming from a career in financial analytics.

These CEOs and company presidents each tailored a business model and plan to their specific company. The model that Paul Lightfoot developed for BrightFarms is particularly innovative, and by summer of 2012, it had led to $60 million worth of "long-term" produce sales since he joined the company in 2010.[6] The strategy is

$2.2 million hydroponic greenhouse,
Lufa Farms, QC

simple: BrightFarms enters into ten-year purchase agreements with supermarket clients, who commit to buying all of the produce grown on the roof (or in a nearby greenhouse facility) at fixed prices. The retailers pay for the produce alone, and not for the greenhouse infrastructure or operations. According to Lightfoot, this model appeals to supermarkets because they consistently receive fresh delicious products at fixed rates, which can provide a competitive advantage over other retailers.[7] Furthermore, fluctuating fuel costs that impact the costs of imported agricultural goods will not sway the fixed prices offered by BrightFarms. In turn, BrightFarms benefits by earning predictable revenue and minimizes risk by contracting with experienced third-party farmers who operate each facility and contractually guarantee produce quality and volume. The strategy pulls from the solar industry, which realized that building owners are more likely to pay for long-term contracts for electricity rather than pay for the rooftop infrastructure itself.

profits + payback

Profit margins vary significantly with production strategy, as do upfront and long-term costs.

Profit margins vary significantly with production strategy, as do upfront and long-term costs. In very general terms, companies that deploy high-cost, high-tech strategies often enjoy higher yields, thereby making money more quickly than low-cost, low-tech rooftop farming companies. Hydroponic greenhouse operations, for example, require a phenomenal upfront investment, but they can also produce high predictable yields year-round that provide high profits. In contrast, other rooftop production strategies generally require a lower upfront investment and garner relatively lower profits.

The 2012 Urban Agriculture Summit, an international conference hosted in Toronto by Food Share and Green Roofs for Healthy Cities, placed Lufa Farms cofounder Kurt D. Lynn on a panel with Brooklyn Grange president and head farmer Ben Flanner to discuss commercial rooftop agriculture. Lynn described the run rate revenue of Lufa Farms' first hydroponic greenhouse facility as $1.3 million per year. With a $2.2 million construction cost (roughly $71 per square foot), and annual yields of approximately 250,000 pounds of produce,[8] Lufa Farms founder and president Mohamed Hage estimates that the payback period for each greenhouse location will be only three to five years.[9] In contrast to this high-input/high-output model, Flanner cited the construction cost of Brooklyn Grange's first row farm as only $200,000 (approximately $5 per square foot),[10] with a payback period of ten years.[11] This installation cost falls into the same range as Eagle Street Rooftop Farm in Brooklyn (see Chapter 5), which cost roughly $10 per square foot to install.

Neither model is better than the other; they are simply different. Business objectives, acreage, availability of upfront capital and expertise of the company's founders help to inform which model a business pursues. Climate can also influence the decision, particularly for businesses operating in cold regions of Canada. Which model makes the most sense for your company?

funding opportunities

One of the most important steps in fleshing out your business plan and projecting your company's profits and payback period is planning how to finance the project. Rooftop farms take advantage of many modes of financing, and there is always room for innovative methods of acquiring start-up capital.

Rooftop farming companies that require more substantial upfront capital, such as Gotham Greens and BrightFarms, often rely upon **investors** to get things cooking. Other companies choose to **self-finance**, at least in part, as was the case with Lufa Farms (see Chapter 5), Brooklyn Grange and I Grow My Own Veggies (in Sarasota, FL). Some take out **business or even personal loans**, and a few rely in part on **equity investments**. Rooftop farms and gardens affiliated with restaurants, such as Uncommon Ground (in Chicago, IL; see Chapter 5), or hotels, such as the Fairmont Waterfront (in Vancouver, BC) or the Fairmont Royal York (in Toronto, ON) in Canada, are generally **funded by the parent company**. Brooklyn Grange tapped into perhaps the most original and diverse funding sources of all. The company's first location relied upon financing from equity investments, small personal loans, personal capital, fundraising events and **crowd funding**. This last method involves pooling small donations from many people who support a cause, which, in the case of Brooklyn Grange, was facilitated by the crowd funding website Kickstarter. The site provides an international platform for individuals and groups to publicize projects and collect a predetermined fundraising goal. Brooklyn Grange pursued a different financing approach for its second row farm location, in which the NYC Department of Environmental Protection (DEP) Green Infrastructure Grant Program funded 75% of the project. According to the DEP website, Brooklyn Grange and the Brooklyn Navy Yard (the facility in which the farm is located) will contribute matching funds of $310,000 to supplement the agency's $592,730 grant.[12] Governmental agencies within several US cities, including Philadelphia and Washington DC, have hosted similar grant programs directed at stormwater management. **Grants** offered by city, state and federal agencies should always be considered for rooftop farming initiatives, particularly when the farms will involve a community education or public health component. Keep an eye out for grant opportunities for rooftop greening offered by a city's water department and department of environmental protection.

One of the most important steps ... is planning how to finance the project.

3. location, location, location

While it's *possible* to grow food on rooftops anywhere in North America, the practice may not always be *practical*.

climate

Always keep in mind that, while it's *possible* to grow food on rooftops anywhere in North America, the practice may not always be *practical*. Climate should be a main consideration in determining in which city or region to site your rooftop farms. Growing food in harsh climates often requires expensive production strategies, some of which may be environmentally harmful. Rooftop agriculture in northern and eastern Canada, for example, requires the use of highly insulative heated greenhouses. Energy is required to heat these glass houses during the colder months, which can result in high fossil fuel use and utility bills. Rooftop farming in the southwestern and south central US is also possible, but it requires massive amounts of irrigation (and possibly misting), in an arid geography that's already stressed for water.

Growing food near population centers throughout North America is crucial but must be balanced with the practicalities of production. As a rule of thumb, temperate, tropical, subtropical and semiarid climates are stable environments for rooftop agriculture. Boreal and desert conditions should be avoided whenever possible. With that said, technology continuously progresses to increase the efficiency of greenhouses and irrigation, thereby pushing the limits of what is both possible and practical.

zoning + policy

After pinning down a specific city or region to pursue, carefully review local codes to determine which neighborhoods permit agriculture and which zones allow for direct consumer sales (see Regulator Comparison table later in this chapter). Frequently, a city's zoning code defines where urban agriculture and the sale of agricultural goods are acceptable land uses, the health code specifies which animals are permitted within city limits, and building codes describe certain incentives for rooftop greening, such as floor area ratio (FAR) bonuses. The zoning codes of certain cities, such as Chicago, even list rooftop farming as a specific use in certain zones.

Restrictions vary by city, but oftentimes agriculture is not permitted in heavy manufacturing zones for food safety reasons. If you have your eye set on a building in a zone that does not permit agriculture, try applying for a variance. The office building that Lufa Farms selected for its first rooftop greenhouse was in a zone that did not permit agriculture, and so the company applied for and received a zoning variance.[13] In cities where antiquated zoning codes do not permit agriculture at all, try contacting a local council person who interacts with the city's planning commission. With enough prodding, sensible conversation and, at times, political pressure, you may effectively influence your city's policies.

Snow-covered rooftop farm, Brooklyn Grange flagship farm, NY
PHOTO BY ANASTASIA COLE PLAKIAS, COURTESY OF BROOKLYN GRANGE

A zoning variance made this Montreal farm possible,
Lufa Farms, QC

The concept of lobbying council persons and the planning commission extends beyond zoning reform. Policy makers have the unique ability to create incentive programs that can make rooftop agriculture more financially viable for individuals, groups and companies. The Regulatory Comparison table that follows highlights progressive policies and incentives implemented in several North American cities that hold great potential for widespread rooftop agriculture (see Chapter 7). As you can see, the table draws upon bylaws that promote green roof construction; regulations that support periphery activities such as urban composting; green infrastructure tax credits, grant programs, expedited permitting; green roof FAR bonuses; green roof tax abatements and incentives for rooftop greenhouses. Other North American cities offer similar policies and incentive programs that promote rooftop

agriculture, such as Seattle, Los Angeles, Minneapolis and Washington DC.

Federal policies may further foster the viability of rooftop agriculture in your city. For example, the Clean Energy Stimulus and Investment Assurance Act of 2009 amended the Internal Revenue Code to allow for a 30% tax credit for green roof construction on residential and commercial buildings when at least 50% green roof coverage is achieved.[14] Similarly, the Environmental Protection Agency has fined several major US cities for violating the Federal Water Pollution Control Act (commonly called the Clean Water Act) by failing to prevent repeated combined sewer overflows. In response to this federal pressure, many of these cities have implemented or plan on rolling out programs to incentivize the construction of green roofs and other types of stormwater management infrastructure.

regulatory comparison
to keep you legal

relevant policies + incentives

CHICAGO	- Expedited permitting for green roof and LEED certifiable building projects[15] - Nuisance control provisions and development standards to prevent composting from threatening public health[16] - Municipal Code explicitly recognizes rooftop farming as an acceptable use in certain zones[17]
PHILADELPHIA	- Business Income and Receipts Tax (formerly called Business Privilege Tax) credit for 25% of green roof construction costs (up to $100,000) when green roof covers at least 50% of total roof area[20] - Periodic Philadelphia Water Department grants for green infrastructure projects[21]
PORTLAND	- Floor Area Ratio (FAR) bonus allowing developers to build an additional 1 sf of building floor for every 1 sf of green roof without additional permitting[24] - Grant reimbursement of up to $5/sf for green roofs, for reducing stormwater infrastructure[25]
SAN FRANCISCO	- Expedited permitting for all green building projects[28] - City agencies required to support urban agriculture through securing state and federal funding, collecting data, supporting gleaning programs, promoting urban agriculture's job training and employment potential, and ensuring that existing urban farms and gardens are utilized[29]

zones allowing agriculture	permitted farm animals	model policy for
- C1-C3, DC, DX, DR, DS, M1-M3, PMD - B3 (community gardens only, with special use approval)[18]	- Honey bees (max. 5 hives, accessory use only, with registration) - Fish (indoors only, with zoning review and building permit) - Horses (with a license) - Chickens, foul (for eggs only)[19]	- Green roof bylaw - Composting regulation
- All zones (community gardens) - All zones except RSD-1-3 (urban farms)[22]	- European honey bees (with permit) - Horses, donkeys, mules (with license) - No others permitted[23]	- Green roof tax credit - Green infrastructure grants
- Open space, employment, industrial, and farm/forest zones - Commercial and single-dwelling zones (conditional use)[26]	- No more than 3 animals total - Chickens, ducks, pygmy goats, rabbits - Turkeys, geese, cows, horses, burros, sheep, llamas, bees (with permit)[27]	- Green roof FAR bonus - Green roof grant reimbursement
- All residential, commercial, industrial, neighborhood commercial and mixed-use districts[30]	- No more than 4 animals total, no more than 3 of a given species, no commercial use - Rabbits, hares, chickens (roosters permitted), geese, ducks, game birds, pot bellied pigs, pygmy goats - Donkeys, cows, goats (with permit)[31]	- Expedited permitting - Green building ordinance - Urban Agriculture ordinance

Zoning codes and policies change periodically.
This chart reflects current information as of August 2012.
Always confirm local codes and policies.

relevant policies + incentives

NEW YORK CITY
- One-year tax abatement of up to $100,000 (or $4.50/sf) for green roofs that cover at least 50% of available roof space[32]
- Periodic NYC Department of Environmental Protection grants for green infrastructure projects[33]
- Rooftop greenhouses can exceed building height limitations when building contains no sleeping accommodations, greenhouse is accessory to a community facility use, 6' setback is respected, and rainwater collection and re-use is incorporated (2012 Building Code Amendment)[34]

VANCOUVER
- Green roof required on all new commercial and industrial buildings over 53,800 sf [38]
- Vancouver Food Charter lays out city government's commitment to promoting local food security through policy and community engagement[39]
- Metro Vancouver Regional Food System Strategy to expand commercial food production in urban areas [40]
- Urban Agriculture Guidelines recommend community garden standards[41]

TORONTO
- Construction of green roofs required on all new buildings with a Gross Floor Area over 2,000 square meters[45]
- Food and Farming Action Plan established goals for increasing capacity for agriculture by 2021[46]

MONTREAL
- Permanent agricultural zones established within city to protect farmland.[49]

zones allowing agriculture	permitted farm animals	model policy for
- R1-R10 (limited use) - C1-C6 (not cited as unacceptable use) - M1[35]	- European honey bees (with permit)[36] - Chickens, rabbits (with permit, none granted on-site with multiple dwellings, coop must be 25' from other occupied homes, no roosters)[37]	- Green roof tax abatement - Green infrastructure grants - Commercial greenhouses - Rainwater harvesting
- RA-1 (limited use) - RS1-7, RT1-10, RM1-6, C1-8 (not cited as unacceptable use) - ALR[42]	- European honey bees (max. 2 hives on residential lots, 4 hives on lots over 10,000 sf)[43] - Hens (older than 4 months, max. 4 per lot, eggs cannot be used commercially) - Rabbits [44]	- Green roof requirements for new construction
- UT (market gardens) - O (gardens acceptable with limited use, not for commercial sale)[47]	- Rabbits (max. 6) - Goats, geese, chickens, turkeys (on land zoned for agriculture only)[48]	- Green roof bylaw
- All zones (community garden)[50] - A1 (agriculture without livestock) - A2 (agriculture with livestock)[51]	- Chickens (for educational purposes in A2 only) - Sheep, goats, horses (A2 only) - Honey bees (A1 only)[52]	- Agricultural zoning

Zoning codes and policies change periodically.
This chart reflects current information as of August 2012.
Always confirm local codes and policies.

proximity to "eaters"

After evaluating local zoning codes and relevant policies, make a shortlist of densely populated neighborhoods that you think would benefit from local food production. Neighborhoods can benefit from local food for different reasons, and so be sure that your company's business approach aligns with neighborhood demands. There may be a market for high-end local food outlets in certain neighborhoods, for which rooftop farms paired with restaurants or gourmet markets would be appropriate. Other neighborhoods may simply lack access to fresh food and would therefore benefit from affordable, nutritious produce through a charitable organization or rooftop CSA. Lufa Farms successfully identified a Montreal neighborhood that lacked fresh produce outlets, and now customers are lined up to buy food basket memberships from the farm (see Chapter 5). Demand for the produce is so high, in fact, that Lufa Farms is opening several new locations within the next few years.

Also keep in mind that transporting food across a city can absorb time and money, so keep it local — *really* local. Try selecting buildings in close proximity to prospective sales outlets. Brooklyn Grange's carefully selected locations allow the company to distribute its food within a five-mile radius. Lufa Farms' first

> **Be sure that your company's business approach aligns with neighborhood demands.**

> **Transporting food across a city can absorb time and money.**

location similarly operates in a ten-mile radius. The only food more local is that which you grow on your *own* roof!

building stock

Evaluating a prospective neighborhood's building stock is the final step in determining where to site your rooftop agriculture operation. Prospective neighborhoods should contain many large (at least 3,000-square-foot) flat-roofed buildings. Ideally, these buildings were built of high-quality steel and concrete, during a period in which local building codes required a high "dead load capacity" for the roof. These criteria may sound specific, but most older cities within the US contain neighborhoods that fit the bill. Consult with several local structural engineers to determine the time period that produced the strongest buildings in your city. In several cities, such as New York City, this time period spans the 1900s to 1950s.

Other building stock considerations involve typical parapet heights, rooftop access features (e.g., headhouses, pilot houses) and building height. If all this seems overwhelming, don't worry! The Decision Tree on the following pages will help guide you through which specific buildings to pursue and what type of agriculture to assume for each.

View of San Francisco's flat-roofed buildings
from Graze the Roof, CA

4. decision tree

Once you pin down a neighborhood in which to site your rooftop farms, try using the **Decision Tree** to quickly assess which production methods (from Chapter 3) are most suitable for each building. This tool can be instrumental in quickly evaluating multiple building sites, and in understanding how multiple farms and varying building types can function as a whole. In addition to aiding business owners, the Decision Tree can also provide city planners and developers with a tool for calculating how much food can be grown in a specific neighborhood or city. For technologically minded users, the Tree can be plugged into Geographic Imaging System (**GIS**) maps through Python programming, in order to map out potential food production patterns in a given area.

The Decision Tree systematically processes several key building characteristics, and recommends between zero and four of the following production methods for each building: containers, raised beds, row farming and/or hydroponics. The Tree takes into account a building's structural integrity, size, ownership, zoning, building use and location. The Tree favors buildings with high **dead load capacity** (as determined ultimately by a structural engineer); **sizable acreage**; building **ownership** or low rental fees; **zoning** that lists agriculture as an acceptable use; **building use** that involves education, food retail or service, or multi-unit occupancy; and a **location** that is near consumers and/or in an area where significant production can occur (e.g., a former manufacturing zone with large unoccupied roofs).

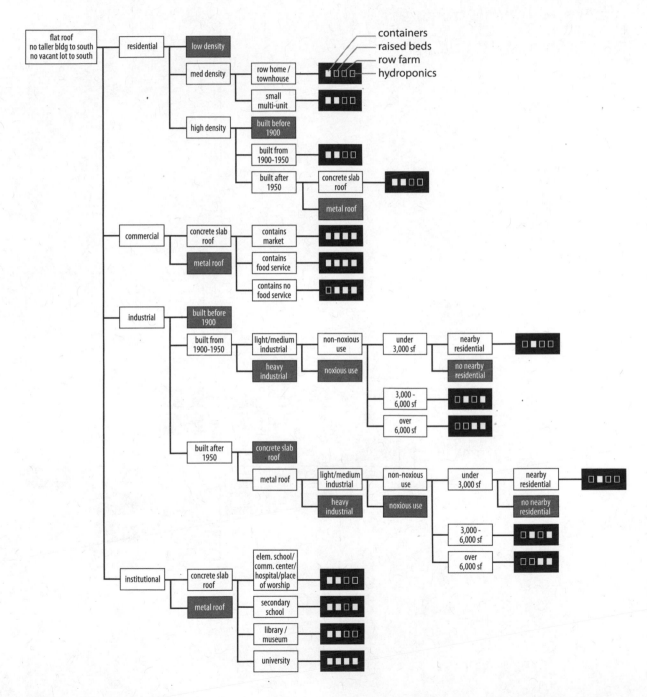

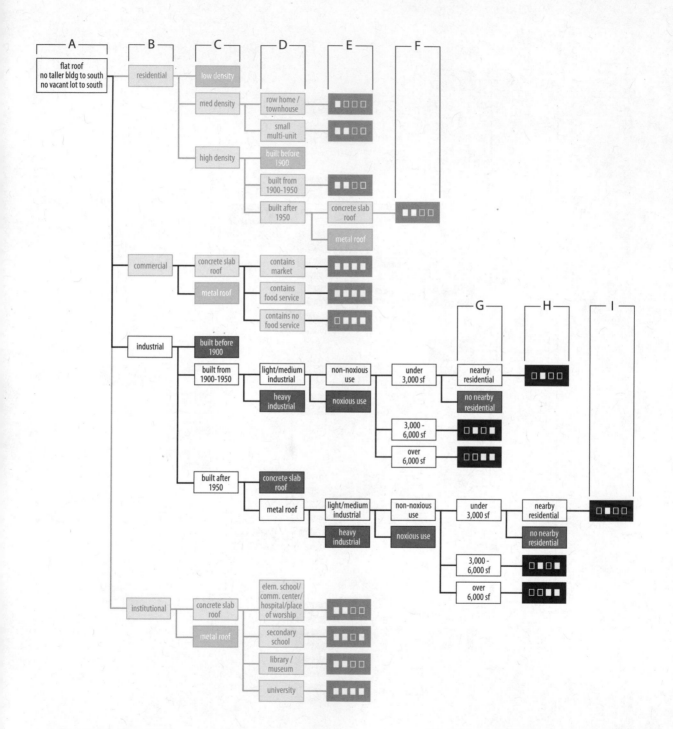

using the tool
decision tree in action

The Decision Tree is read from left to right, with various building conditions listed in columns. In order to determine if it is suitable for rooftop agriculture, the building must first meet the preliminary requirements, which appear in column A. If these preliminary conditions are met, then the first **decision** comes under column B of the Tree. This decision addresses the building's zoning, whereby either "residential," "commercial," "industrial" or "institutional" is selected. According to the Decision Tree, all zoning categories are acceptable for rooftop agriculture.

The options for decision C are dependent upon the answer to decision B. In other words, if "residential" was selected for decision B, then decision C relates to building density. If "commercial" was selected for decision B, then decision C addresses roofing material. This same structure applies to the rest of the Decision Tree, whereby each building for which the Tree is used follows a linear decision-making path until either a **"go"** or **"no go"** cell is reached.

The **"no go"** cells throughout the Decision Tree indicate when a building is most likely not suitable for rooftop agriculture. This outcome can arise as a result of inappropriate zoning, structural, locational or other related variables. In contrast, the **"go"** cells indicate when a roof is a viable candidate for rooftop production.

Within each "go" cell is a recommendation for which types of agriculture are most suitable for the building. The recommendation appears as four white boxes, which correspond to the agricultural methodologies in Chapter 3: containers, raised beds, row farm, and hydroponics. In the Decision Tree, recommended methodologies appear as solid white boxes, and an empty box denotes a methodology that is not recommended. The "go" cell in column F, for example, recommends containers and raised beds.

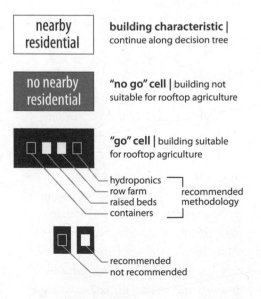

5. forging relationships

universities + research institutions

Working collaboratively with universities and research institutions can prove fruitful.

Relationship building is key to the success of any rooftop farming operation. Working collaboratively with universities and research institutions can prove fruitful in terms of research development, and potentially with grant-funded initiatives. The idea of research partnerships is nothing new for the green roof industry. For example, the company I work for designed and manages highly monitored vegetated roofs, or "research green roofs," as they're called, at St. Joseph's University (in PA), Hillside Elementary School (in NJ), Walmart Store No. 5402 (in IL) and Walmart Store No. 5899 (in OR). Various types of researchers collect data from each of these roofs, as you may imagine, but the sites all produce useful data and promote ongoing collaboration.

In Montreal, the Rooftop Garden Project built an educational rooftop garden at McGill University, with the help of students. The garden demonstrates the potential of rooftop agriculture, while providing fresh produce to students, faculty and staff. Toronto's Ryerson University is considering an on-campus rooftop vegetable garden for similar educational and demonstration purposes. Other North American universities that have invested in researching vegetated rooftops, and potentially collaborating with rooftop farming companies, include Penn State University, the University of Maryland and Michigan State University.

distributors + food aid

Relationships with food distributors and food aid organizations can further the success of your company, and ensure that the produce reaches your target audience. In New York City, Gotham Greens forged profitable relationships with several high-end grocery chains, including Whole Foods Market. BrightFarms, which designs and manages hydroponic greenhouses throughout North America, similarly develops long-term sales relationships with supermarkets. Eagle Street Rooftop Farm (in Brooklyn, NY) distributes to several restaurant locations, and must continuously maintain and build new relationships.

When it comes to food aid, Graze the Roof takes the cake. This San Francisco garden sits atop a church, with robust outreach, public assistance and after-school programs, and functions in direct partnership with the church (see Chapter 4). The garden feeds those in need, and those who benefit from healthy eating activities. SHARE Food Program, a Philadelphia-based non-profit food distribution center and ground-level urban farm, provides affordable food packages to those in need. The packages are supplemented with farm-fresh produce, and sometimes with rooftop honey produced on location by Urban Apiaries (see Chapter 5). SHARE's long-term vision is to farm its warehouse building's roof, in order to further supplement the food packages.

community education + outreach

In addition to supporting food aid assistance, Graze the Roof serves to educate school groups and community members. In fact, education and outreach are so important to the garden that Graze the Roof employs two garden educators (see Chapter 4). Lufa Farms, in Montreal, similarly invests in community education.

The company's Community Team works to educate neighbors about fresh food and urban agriculture, in order to promote healthy eating and hopefully boost sales (see Chapter 5). Whatever the reasoning for investing in education and outreach, companies are beginning to realize the power of community in building support for their gardens and farms. Both non-profit and for-profit rooftop farming companies will continue to build these important relationships with community members.

Community education and outreach with kids, Graze the Roof, CA
PHOTO BY MICHAEL I. MANDEL, COURTESY OF GRAZE THE ROOF

checklist, check!

Now that you're acquainted with the considerations developing a rooftop farming company with multiple farm locations, try filling out the **Industry Checklist**.

How do you rank? Is your business objective clear and defined? Have you strategized how to finance your company? Are there universities and organizations with which you can build relationships?

If you have trouble checking off any of these boxes, consider hiring a consultant or business advisor. If business is not your strong suit, and you foresee business planning and development as legitimate obstacles to your company's long-term success, you may want to think seriously about taking on a business partner. Many companies rely upon leadership with varying areas of expertise, and commercial farming on rooftops requires different skill sets than ground-level production. Other partners that may help you complete the Industry Checklist include volunteer coordinators, community outreach specialists and researchers.

☐ **Primary objective**

    ☐ Underlying purpose of initiative determined

    ☐ Long-term goals identified

    ☐ Milestones and measures of success mapped

☐ **Financing**

    ☐ Triple bottom line considered

    ☐ Business plan clearly defined

    ☐ Profits and payback period understood

    ☐ Funding opportunities reviewed and understood

☐ **Location, location, location**

    ☐ Climate limitations understood

    ☐ Local zoning and policies understood

    ☐ Proximity to "eaters" evaluated

    ☐ Building stock assessed

☐ **Decision Tree**

    ☐ Tree consulted when selecting production methods

☐ **Forging relationships**

    ☐ Universities and research institutions contacted

    ☐ Partnerships with distributors and food aid entities explored

    ☐ Community education and outreach planned

7 | Potential Hotspots

A handful of North American cities are ripe for rooftop agriculture, and they're ripe now. In the US, Chicago, Philadelphia, Portland, San Francisco and New York City all contain the physical infrastructure and cultural drive necessary to support rooftop agriculture. In Canada, Vancouver, Toronto and Montreal all possess successful pockets of rooftop farms and gardens, and the policies in place to foster more.

This chapter explores the defining urban characteristics that create a canvas fit for rooftop agriculture. Physical conditions such as building density and abundance of large building footprints are considered, as are cultural qualities like socioeconomic patterns, existing food streams, food culture and degree of local activism.

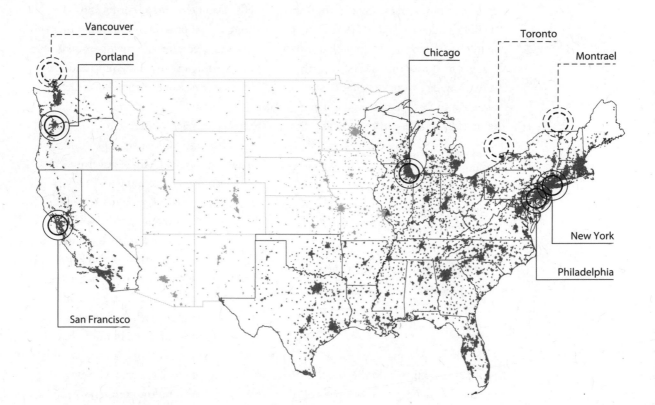

Vancouver

Portland

Chicago

Toronto

Montrael

San Francisco

New York

Philadelphia

what makes a hotspot?

Rooftop agricultural hotspots are cities or specific neighborhoods that contain a relative abundance of rooftop farms and gardens. With three of the nation's leading rooftop farms and dozens of rooftop vegetable gardens, Brooklyn, NY, is the top dog in today's growing boutique industry. We'll get to Brooklyn's specifics shortly, but for now, let's focus on what conditions must align to foster a potential hotspot. The most basic requirements are as follows:

1. **Climate** able to sustain a lengthy growing season (at least 8 months)
2. **Government** that supports, or does not prohibit, urban agriculture
3. **Building density** that is medium to high by North American standards
4. **Building stock** with widespread flat roofs that can sustain heavy loads

Generally speaking, the North American cities that hold the greatest potential for widespread rooftop agriculture are either **dense metropolises** in need of local food production (like New York City) or urban areas that contain a **robust food culture** and **socially active residents** (like San Francisco). The former category speaks to need, while the latter resonates with desire. Both need and desire are powerful forces with a high propensity to elicit action, so these cities are primed for entrepreneurs and everyday gardeners to get up there and get started.

Secondary characteristics to keep in mind when evaluating your city include high customer demand for local food, a strong neighborhood or citywide gardening culture, community members dedicated to addressing urban food equity and the presence of research institutions (such as universities with agricultural or horticultural departments) interested in developing new urban agricultural strategies and furthering the field. Cities' governments that incentivize entrepreneurial activity are also beneficial, as are cities that offer tax incentives or reduced stormwater fees for green roofs. An abundance of buildings with **large footprints** is also key, particularly when promoting widespread commercial agriculture. These structures often include industrial, institutional and office buildings, but keep in mind that networking smaller roofs can achieve the same scale as one large roof. Cities that place high value on the community aspect of rooftop agriculture may benefit from the presence of **mixed-use** neighborhoods that permit agriculture. This type of development allows rooftop farms and gardens to exist right near where people live, rather than off in another part of town.

Let's see how North America's potential hotspots rank.

Dense metropolises in need of local food ... or urban areas that contain a robust roof culture and socially active residents,

Heavily branded Brooklyn Grange honey jars,
Farmers' Market, NY

Chicago, IL

America's third-largest city knows how to grow food, and knows how to encourage city government to implement policy that makes urban farming and gardening even easier. For years, Chicagoans have grown food in vacant lots, home gardens and community gardens. In an effort to promote commercial agriculture within the city, Mayor and former White House Chief of Staff Rahm Emanuel worked with city agencies, urban agriculture experts and community members to forge new policies. In 2011, Chicago's city council successfully approved a zoning code amendment that increased the allowable size of agricultural plots within city limits to 25,000 square feet, thereby accommodating commercial urban farms. Additional policy changes now permit the limited sale of agricultural goods within residential neighborhoods, allowing urban farmers to grow and sell food near their customers.

Interestingly, the city government also recognizes the value of **green roofs**, which are installed throughout the city, even on City Hall! Buildings that contain green roofs are subject to lower municipal stormwater fees, and so the owners of large and small buildings are well acquainted with this rooftop option. The regular occurrence of vegetated roofs throughout the city suggests that Chicagoans already embrace the concept of utilizing roof space, and may not think rooftop agriculture is so far-fetched. For some, growing food on rooftops may even seem like the logical next step. An experimental rooftop farm in Chicago's Back of the Yards neighborhood is one of the first to take the step. Farmed by Urban Canopy, The Plant is gaining attention and breeding innovation.

Chicago's **local food culture** further elevates the city as a potential hotbed for rooftop production. The popularity of restaurants like Uncommon Ground (see Chapter 5) attest to Chicago foodies' love for local organic cuisine and their drive to support businesses that grow food on location. With politically charged residents, green roofs galore and a strong local food culture, Chicago may just be one of the next big rooftop agricultural hotspots. Let's make it happen!

Chicago

Philadelphia, PA

Philly is hot for urban agriculture. The farmers here, in my hometown, are young, energized and full of fresh ideas. The growing agricultural community coincides with the city's rising demand for local food, and rooftop agriculture is all the buzz. Philadelphians want their farmers' markets, they want their CSA shares, and they want fresh locally grown food on their plates in restaurants. Many of the city's trendy BYO restaurants, in fact, pride themselves on local farm-fresh fare.

Urban agriculture in Philadelphia is not just a priority of the upper crust; many impoverished neighborhoods survive off food from local farms and gardens. Sections of West and North Philadelphia, in particular, grow food in community garden plots within urban farms like Mill Creek Farm and Aspen Farms. So much vacant land has been converted into urban farms and gardens that University of Pennsylvania researcher Domenic Vitiello has mapped and recorded data from these gardens for years.

Philadelphia's widespread food inequity inspired a food aid distribution center in North Philadelphia, called SHARE Food Program, to establish a farm on its premises. This farm, which provides fresh food to local Philadelphians in need, has become so popular that SHARE hopes to convert 160,000 square feet of its warehouse rooftop into the city's first commercial-scale rooftop farm. With successful, existing rooftop agriculture operations such as the Sand and Alarcon Residences (see Chapter 4) and Urban Apiaries (see Chapter 5), Philadelphia already has one toe in the swimming pool. A commercial farm atop SHARE may be enough to commit the city to a swim session. In 2012, the City Planning Commission released a new zoning code that permits community gardening in all zones and "Market or Community-Supported Farms" in all zones except for low-density residential zones (RS-1, RS-2, and RS-3).[1] These code updates could provide a springboard for urban agriculture on the ground and on rooftops.

Philadelphia
IMAGE ©2012
TERRAMETRICS, OBTAINED
FROM GOOGLE EARTH

Above:
The hotel's 2,100-square-foot roof
garden, The Fairmont Waterfront, BC

Right:
Executive Chef Dana Hauser
The Fairmont Waterfront, BC
PHOTOS BY AND COURTESY OF
FAIRMONT HOTEL & RESORTS

Above:
Project Managers + Garden Educators
Nikolaus Dyer (left) and Lindsey Goldberg
(right), Graze the Roof, CA

Right:
Nikolaus Dyer tending to alpine
strawberries, Graze the Roof, CA
PHOTOS BY MICHAEL I. MANDEL,
COURTESY OF GRAZE THE ROOF

Brooklyn Grange Head Farmer Ben Flanner working a farmers' market stand, Brooklyn Grange Brooklyn Navy Yard Farm, NY

Top: Mounded crop rows in full production, Eagle Street Rooftop Farm, NY

Right: Rooftop pepper harvest and brand label for value-added products, Eagle Street Rooftop Farm, NY

The children enjoy helping out in the Sand family's rooftop garden

Home gardener Jay Sand and his three daughters, Sand Residence, PA

Lufa Farms sits atop a fully functioning office building, Lufa Farms, QC

Peppers from the 31,000 square foot greenhouse, Lufa Farms, QC
Photos by and courtesy of Lufa Farms

Left: Terra Fata Farm, CO
PHOTO BY TRENTON BARNES,
COURTESY OF EMILY HARTNETT

*Right: Blushing Goat
Farm, CO*
PHOTO COURTESY OF
LEAH CAPEZIO

*Below:
Grazing rooftop goats,
Al Johnson's Swedish
Restaurant + Butik, WI*
PHOTO BY MATT NORMAN
PHOTOGRAPHY, COURTESY OF
AL JOHNSON'S SWEDISH
RESTAURANT + BUTIK

Urban Apiaries operates on the roof of Philadelphia's SHARE Food Program, PA

Jay Sand interview
Philadelphia Rooftop Farm (PRooF), President

LM Do you think that large-scale rooftop agriculture is viable in Philadelphia?

JS Yes absolutely. Whether it's actually going to happen is another question, but it's possible.

LM Can rooftop agriculture succeed as both a community building tool and a commercial enterprise?

JS Absolutely both. It has to be a community project because it requires community support to make it happen, especially in residential neighborhoods. Neighbors are very important in the success of rooftop agriculture, and in commercial areas, fellow businesses are the neighbors .... Commercial production [on industrial roofs] is absolutely viable. Farming in industrial areas is a great way to use the space .... Regardless of location, **bounty of food leads to the sharing of food. That is a great community building tool.**

LM How viable would large-scale rooftop agriculture be in New York City?

JS New York is crazy because it's so big, but absolutely .... It's different from Philadelphia because it's so massive. When things happen in one part of [Philadelphia] people catch wind of it .... There's a lot of communication within neighborhoods and between neighborhoods. There's no reason why it couldn't work; New York would be a fantastic place .... The mindset here [in the US] has always been that there's more room. But what if there's not?

LM Do you think that rooftop farming is viable in the US?

JS Yes. In the places where the infrastructure is friendly to it. There's so much room in the US that it would serve as a very tiny part of commercial agriculture. But there are **parts of the country where people need access to food,** and it could be very useful in those urban areas.

LM Is rooftop agriculture a fad that may pass?

JS If food were a fad then people would have stopped eating it a long time ago.

"The mindset here [in the US] has always been that there's more room. But what if there's not?"

"If food were a fad then people would have stopped eating it a long time ago."

Portland, OR

Portland's city government leads the nation in progressive environmental policy, and often lays the sustainability pathway for others to follow. With a history of stormwater management incentives and progressive urban growth policies, Portland is no stranger to being first. A classic example of the city's **progressive policy** is its green roof incentive program, which reduces municipal stormwater fees by up to 35% or grants a floor area ratio (FAR) bonus to building owners who install a green roof.[2] In contrast to Chicago's bottom-up approach to policy changes, Portland tends to exhibit a top-down approach, where the change implemented by policy makers soaks through the rest of the community. Portland's government now has the opportunity to pioneer policies and incentive programs that promote rooftop agriculture as an affordable stormwater management and food localization tool.

In addition to leading the country with progressive policy, Portland was one of the first **local organic food culture** trendsetters.

Having spent much of 2008 in Portland, I found it remarkably difficult to find restaurants that did *not* offer organic food. Health-food stores are abundant throughout the city, as are home, community and university campus gardens. Portland's flat-roofed building stock furthers the city's potential for widespread rooftop agriculture. Industrial buildings and even downtown lofts in the Pearl District could be suitable locations for rooftop farms and gardens. A small handful already exists within the city, such as that atop Noble Rot, a downtown restaurant that features local organic food. Most of Portland's rooftop gardens are relatively small and experimental, but the cool, wet climate begs for larger rooftop farms. Its abundant gardening enthusiasts and activists could easily provide the expertise and enthusiasm necessary to take this initiative off the ground. Truthfully, with so many favorable characteristics, it's strange that rooftop farms and gardens aren't already more prevalent throughout the city.

Portland
IMAGE ©2012
TERRAMETRICS, OBTAINED
FROM GOOGLE EARTH

San Francisco, CA

As the birthplace of Alice Waters' local organic food movement, the Bay Area has possessed a **robust food culture since the 1970s.** Within the region, San Francisco in particular has the ability to become an agricultural hotspot, given its reputation as a **socially active hub** and its relatively high population density. From an architectural perspective, San Francisco's building stock is dominated by **flat roofs,** which provide ideal infrastructure for rooftop agriculture. Successful agricultural roofs such those at Graze the Roof (see Chapter 4), the Chronicle Building and the Fairmont Hotel San Francisco are beginning to emerge around the city as a testament to these favorable conditions.

According to a rooftop resource study published by the local organization Bay Localize, a "rooftop resource development philosophy is emerging and taking root in the Bay Area."[3] The organization suggests that not only does notable **community interest** exist in using roofs for food production and community empowerment, but building owners and developers are looking to rooftop vegetation in an effort to maximize building functioning. While local and state governments do not currently provide financial incentives for rooftop agriculture in the Bay Area, government action (or in this case, *re*-action) will ultimately follow the community and developers' lead. This is one hot area to keep your eye on.

San Francisco
IMAGE ©2012
TERRAMETRICS, OBTAINED
FROM GOOGLE EARTH

New York City, NY

New York City
IMAGE ©2012
TERRAMETRICS, OBTAINED
FROM GOOGLE EARTH

The constellation of rooftop farms and gardens dotted across New York City's skyline position this metropolis as North America's rooftop agriculture poster child. With an incomparable **population density** of over 27,000 people per square mile,[4] New York City has a whole lot of mouths to feed. Where does all this food come from? Well, if it's not grown on a roof or ground-level community garden, then it's imported. Food may arrive from a farm in New Jersey or Upstate New York, or it may travel from another country around the globe. Either way, planes and trucks and traffic are required to get this food to your plate. A major argument for building more rooftop farms and gardens throughout the city is to decrease New York's import-dependence, traffic congestion and air pollution. Keep it local, keep it clean.

Despite what you may think, New York City's climate lends itself to growing crops. Due to a condition known as the urban heat island effect, the city is actually hotter than surrounding areas. This added heat (which is generally viewed as a deleterious phenomenon) can benefit urban agriculture by extending the **growing season**. Production time will never be as long as that in California, but the extended growing season does give New York City an advantage over other Northeastern and Midwestern cities.

New York City's **capital wealth** also lends itself to the proliferation of rooftop agriculture. Since the initial cost of installing a rooftop farm or garden is generally higher than that on the ground, external funding is often necessary to get a project started. Several rooftop farms in New York City were founded with the generosity of investors and other capital campaigns, and may not have sprouted up without this assistance. And sprout they do! In Brooklyn, Long Island City, the Bronx, Manhattan and new locations every day. Brooklyn's Greenpoint neighborhood is particularly attractive to rooftop farmers and gardeners, so keep a lookout for new rooftop additions in this veritable hotspot.

From a global perspective, New York City's reputation as a **showcase for the** **spectacular** places the metropolis in a position to become an international leader in rooftop agriculture. The ability for **new trends** to emerge out of the city has proven successful for generations, extending from fashion to art to, most recently, landscapes. Manhattan's new High Line and East Side Waterfront Park are two examples of how the city is stretching the boundaries of public space and reuse in order to produce innovative functional places that expand the status quo and inspire other cities to follow suit. The adoption of a cohesive rooftop agricultural strategy at the neighborhood, borough or city scale would propel New York City's rooftop agriculture industry forward, and could stimulate other urban centers around the work to assume similar strategies.

Eagle Street Rooftop Farm, NY

Vancouver, BC

Vancouver

In Vancouver, British Columbia, three hours due north of Seattle, green roofs, terrace planters and even trees peak out from building tops along the city's streets. What about rooftop vegetable gardens and farms? Well, a handful of businesses, community centers and residents are trying their hand at that too. From The Fairmont Waterfront hotel, to YWCA Metro Vancouver charity, to rooftop plots on apartment and condominium buildings, vegetables are ripening across the skyline.

Unlike most Canadian cities, Vancouver has a mild climate that allows for year-round food production. The Pacific Ocean moderates the city's temperatures, which rarely reach above 70°F in the summer and below 32°F in the winter. Like Seattle, the city sees steady rain during the winter months, which provides plenty of water for thirsty fruits and vegetables. And then there's the local culture. Many Vancouver residents embrace gardening, local food, organics and just about anything deemed sustainable. This green mindset makes the argument for rooftop agriculture within the city that much easier. Pair that with progressive city planning policies and environmentally minded local and federal governments and *voilà*. Let's cross our fingers for more rooftop farms and gardens germinating across the city!

Toronto, ON

Toronto has more mouths to feed than any other metropolis north of the border.[5] While the city continues to grow and subsume its surrounding farmland, green roof advocates are thinking seriously about rooftop acreage. The international green roof trade organization, Green Roofs for Healthy Cities (which happens to be based in Toronto), spurred public policy graduate students from the University of Toronto to put a price tag on the benefits of widespread rooftop agriculture within the city. While the 2009 study is not peer-reviewed, it reaches some interesting conclusions regarding the monetized benefits of food production on 12,000 acres of urban rooftops. The study found the expected dollar value return of a 1,000-square-foot rooftop production area to be just over $450, which equates to an annual return of $1.7 billion for the City of Toronto.[6] The study also estimated that Toronto imports 32,000 tons of produce each year from the US alone, which yields over 16,000 tons of transportation-related carbon dioxide emissions per year.[7] While the exact figures cannot be confirmed without additional research and peer-reviewed studies, the message is clear: growing food locally saves people money and cuts down on greenhouse gas emissions. It also feels good.

Greening rooftops is nothing new for Toronto. In 2009, it became the first North American city to enact a bylaw that requires and governs the construction of green roofs on all new buildings with a gross floor area over 21,500 square feet.[8] The bylaw has been hugely successful in promoting green roofs, and widespread rooftop agriculture is the next logical step.

While vegetable gardens are populating rooftops across the city, one of the most notable gardens occupies the roof of the Fairmont Royal York hotel. Like its counterpart in Vancouver, this roof garden supplies the hotel's kitchen with roof-fresh fruit, vegetables and herbs. Of more historical significance, perhaps, is Annex Organic, Toronto's very first rooftop farm, founded in 1996.

Toronto
©2012 Cnes/Spot Image, Image ©2012 DigitalGlobe, Image ©2012 Digital Globe, Image ©2012 TerraMetrics, obtained from Google Earth

Montreal, QC

Two hours due north of Burlington, VT, Canada's second most populous city[9] rests at the confluence of the Saint Lawrence and Ottawa rivers. Montreal experiences warm summers (rarely reaching 80°F) and harsh snowy winters, which makes year-round food production a challenge. As farmland evaporates into housing developments, greenhouses and plastic hoop houses may be the ticket to widespread urban agriculture across this city's acres of flat roofs.

At least one Montreal company has picked up on the awesome potential for rooftop greenhouses. Since 2011, Lufa Farms has provided fresh vegetables and herbs to roughly 2,000 people every week! The farm's 31,000-square-foot hydroponic greenhouse rests atop a fully functioning office building, which directly plays into one of the company's tag lines: "More mouths to feed. Less land." At a smaller scale, the Rooftop Garden Project, founded in 2003 by the international development organization Alternatives, acts as a resource for gardeners who want to grow food on underutilized surfaces throughout the city. It developed a low-cost, lightweight self-watering container built from everyday materials. The Rooftop Garden Project releases DIY guides that teach home gardeners and school groups how to build containers and edible rooftop gardens.

In our 2010 interview, the Rooftop Garden Project's project coordinator Ismael Hautecœur explained that the organization's approach involved developing visible, strategically located "showcase gardens" to demonstrate ease and importance or rooftop gardening. To date, the Rooftop Garden Project has inspired children, adults and even senior citizens to garden rooftops, balconies and plazas across the city, including McGill University. Although the city's zoning code allows for community gardens in all areas, it permits commercial agriculture and the keeping of farm animals only in designated agricultural zones. Rooftop farms in non-designated zones must request a zoning variance, as did Lufa Farms.

Montreal
©2012 Cnes/Spot Image, Image ©2012 DigitalGlobe, Image ©2012 GeoEye, obtained from Google Earth

Ismael Hautecœur interview
The Rooftop Garden Project, Project Coordinator

LM What is the Rooftop Garden Project's philosophy?

IH A big part of our philosophy is to inspire initiatives within communities. We start by developing large strategically located "showcase gardens," such as our garden at McGill [University]. Our strategy also includes making rooftop agriculture seem like a necessity; convincing people that all of its benefits are important, not just one or two; and working bottom-up and top-down simultaneously .... It's important to understand that **people, not space, are needed for success,** and that spaces should be intergenerational and inter-cultural. [Rooftop] gardens should be visible, accessible, beautiful and educational.

LM What user groups are present in the gardens?

IH Diverse populations use the gardens. **Youth are dominant** in the Rooftop Garden Project gardens, while women outnumber men .... The gardens are also very popular with certain immigrant communities, especially with Italians, Portuguese and Asians. [Home agriculture] is already a part of their culture ... [and while] most people garden recre-ationally, **some immigrants grow for subsistence.**

LM Is the city of Montreal zoned for agriculture?

IH Agriculture is permitted anywhere, not just where it's zoned. People don't realize this, and so they don't consider growing food in urban areas .... Zoning [specifically] for urban agriculture and peri-urban agriculture would help the situation.

LM How do you select which roofs to garden first in a city?

IH The **roofs should be accessible, structurally sound and secure,** and underused terraces should also be used .... It's also effective to promote [rooftop vegetable gardens] with the developers of new buildings.

LM Why does the Rooftop Garden Project choose to work with universities?

IH **Universities are very powerful and influential.** They are effective places for starting new types of initiatives, and the public has confidence in them. [Universities] also provide high exposure for many types of people, and publicity can easily follow. It's also easier to get press when you're partnered with a big name university or with a city.

> "[Rooftop] gardens should be visible, accessible, beautiful and educational."

regional synthesis

North America is blessed with an abundantly rich geography, climate and population. This diversity demands tailored rooftop agricultural strategies that respond to the unique conditions of each region. The regional synthesis map offers generalized approaches for those North American regions in which rooftop agriculture offers the most promise. Faded back areas are not appropriate for widespread rooftop agriculture, due to low population densities and/or unfavorable geographies.

To generalize, the **West** benefits from rooftop agricultural strategies that emphasize community involvement and water conservation. The region's high urban population and low levels of precipitation feed into this assertion as does the region's propensity to support community initiatives. Also, the abundant rural agriculture that exists near cities within this region suggests that non-commercial rooftop agriculture (gardens) may be more suitable than rooftop farms. Other details of production should be adopted accordingly, as climate, culture and development patterns vary within the Pacific and Mountain divisions of this region (as defined by the US Census Bureau).

Strategies for the **Midwest** should focus on rooftop gardening as a teaching tool and the production of food for consumption by people within the building below. These strategies would not only empower Midwesterners but also provide green jobs and improve nutrition in office buildings, schools and restaurants.

The **Northeast** benefits from high-yield commercial farming that addresses food localization. Emphasizing food security and minimizing the transportation of food should be paramount, given the region's high urban population. In dense cities, incorporating community-oriented rooftop strategies is just as important as producing high yields, given the potential for building community, providing green space and reconnecting children and adults with their food.

Strategies for the **South** should emphasize high-value commercial crops and food production for use within buildings. The South's warm climate and high precipitation levels are ideal for novice and professional growers alike.

The **Canadian cities** highlighted in this chapter benefit from following similar approaches in proximate American cities. Vancouver and Portland, for example, would likely deploy similar strategies to achieve overlapping goals. Small pockets of rooftop agriculture may appear in northern and central Canada, but the country's most densely populated cities hold the most promise.

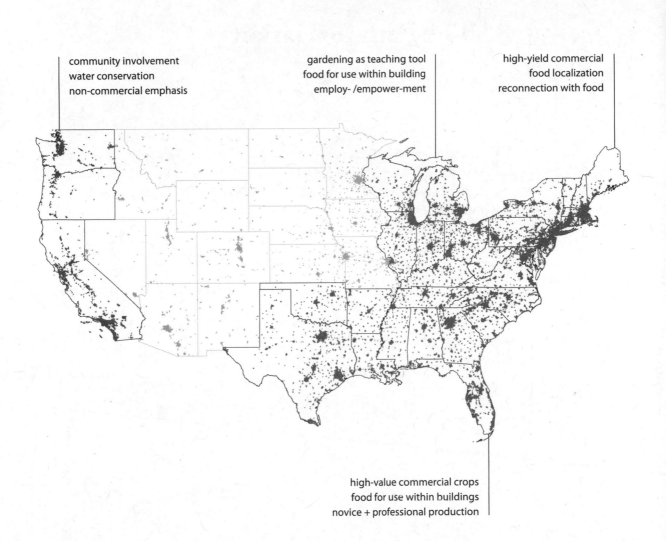

community involvement
water conservation
non-commercial emphasis

gardening as teaching tool
food for use within building
employ- /empower-ment

high-yield commercial
food localization
reconnection with food

high-value commercial crops
food for use within buildings
novice + professional production

8 | The Path to Market

The rooftop agriculture movement is real. It is dynamic. It is blossoming. Every day the movement's energy and excitement infiltrate further into schools, restaurants, office buildings and warehouse spaces — rooftop agriculture is contagious. As I sit here writing this book, the phenomenal progress the movement has experienced during the past few years overwhelms me. Every day more and more people climb up onto rooftops, build farms and gardens and grow community. Policies promoting the success of rooftop agriculture germinate left and right. Academia and the media buzz with analysis and coverage of the movement. It's really happening.

Freshly harvested radishes, Eagle Street Rooftop Farm, NY

the three ps

Rooftop farmer Annie Novak once told me that *people, plants and policy* are "the three Ps of successful rooftop agriculture."[1] And she's right. As people rediscover the taste of fresh food and their passion for community building, demand for rooftop gardening grows. As farmers migrate to the urban frontier to plant seeds on rooftops, their discovery of skyline growing techniques and adaptable plant cultivars expands. As food insecurity becomes a reality for more cities, urban agricultural policies mature. Each P — people, plants and policy — works to further one segment of the rooftop agriculture movement. The movement's long-term success, however, depends upon the three P's operating in synergy.

The movement's success also depends upon diversity. This is not a movement built exclusively of young people or tree huggers or idealists. This is a movement of ordinary people like you and me, from all walks of life, who work together to turn visions of fresh food and green space into reality. Each person invests in rooftop agriculture for a reason as unique as one's own personality and gains something uniquely precious in return. Those digging deeply into rooftop soil rely upon collaborators from diverse backgrounds — engineering, architecture, business, community outreach — to make their skyline farms and gardens tick. Success requires young leaders with foresight, older mentors with hindsight and a village of supporters all around, as demonstrated by the skyline farms and gardens throughout this book. Diversity in plant material is equally important in the movement's success, as rooftop agriculture would not flourish without diverse crops that cater to both rooftop survival and taste. Diverse policies and incentive programs also are equally vital in enabling rooftop agriculture's long-term success. As shown in this book, effective top-down strategies target everything from stormwater management to greenhouse design to community gardening. The rooftop agriculture movement's success relies upon the coalescence of diverse people, diverse plants and diverse policy.

Innovation also makes the movement tick. Cutting-edge technology continually bolsters the efficiency of hydroponic greenhouses. Innovative business models reshape the ways in which retailers value produce. Creative educational programs reach urban kids who would otherwise have no access to gardens. Every day new innovations propel the movement forward, and every second, opportunities arise for rethinking our food system.

The [rooftop agriculture] movement's long-term success, however, depends upon the three Ps operating in synergy.

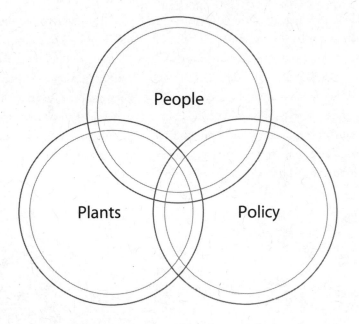

The "three P's" of successful rooftop agriculture

a unique environment

When innovating, attracting others to the movement or harvesting fresh vegetables, always remind yourself that growing food on rooftops is profoundly different than cultivating fruits and vegetables on the ground. Rooftop farmers and gardeners constantly experiment, learn from their mistakes and adapt to the unique skyline environment. A primary goal of this book is to introduce you to the obstacles and opportunities of rooftop agriculture so that you can confidently move forward with your project and succeed. When one roof does not provide enough acreage, create a network of roofs that work together. When unsure how to finance a project, look to strategies that other rooftop farms and gardens pursued.

Young rooftop gardeners hunting for tomatoes,
Sand Residence, PA

you have the tools

This book gives you the tools you need to succeed. Whether you're reading from cover to cover or scanning bits and pieces of relevant material, you now have these resources at your fingertips:

1. **Detailed benefits** of rooftop agriculture, to fuel the case for your farm or garden (Chapter 2);

2. **Production Method Comparison**, to inform your decision of which agricultural strategies to pursue (Chapter 3);

3. **Crop Planting Guide**, to help you select the most appropriate crops (Chapter 3);

4. **Garden Checklist**, to prepare you for success in rooftop gardening (Chapter 4);

5. **Farm Checklist**, to crosscheck the planning and development of your rooftop farm (Chapter 5);

6. **Business Diagram**, to establish your rooftop farming company's long-term goals and milestones (Chapter 6);

7. **Decision Tree**, to help you quickly assess which buildings are most suitable for rooftop agriculture (Chapter 6);

8. **Regulatory Comparison chart**, to introduce you to relevant rules and regulations (Chapter 6);

9. **Industry Checklist**, to anticipate the needs of your multi-roof farming operation (Chapter 6);

10. **Regional Synthesis**, to reveal the most promising rooftop agricultural orientations for each North American region (Chapter 7); and

11. **Insider information** from over 15 of the industry's leading farmers, gardeners, educators, CEOs, landscape architects, chefs and green roof professionals.

You are holding the most comprehensive resource to date on rooftop agriculture. Let the words of well-seasoned skyline farmers like Annie Novak, Dave Snyder and Ben Flanner inspire you. Learn from the business decisions of rooftop farm CEOs like Mohamed Hage and Viraj Puri. Introduce fresh food to your community like Jay Sand, Lindsey Goldberg and Vincent Dessberg. Bring wisdom from your own rooftop farm or garden to the table and share your knowledge with friends, neighbors and fellow enthusiasts around North America. Teach the next generation of rooftop farmers and gardeners how to steward not only sacred spaces on the ground, but also those across our cities' skylines. You have the tools to succeed.

*Farm apprentice
harvesting basil,
Eagle Street Rooftop
Farm, NY*

onward and upward

Since my research for this book began
in 2009, rooftop agriculture has matured
from a boutique industry into an integral
facet of the urban food system. Urbanites
are planting seeds, buying local food and
demanding legislation that forwards their
cause. We are successfully rethinking the
avenues that food takes from seed to urban
plate, while nourishing both our stomachs
and our souls.

You have the tools you need to join
this movement and propel it even further.
Now go do it. Grab a spade, grab a fork,
grab a pen — go out and get started.

Glossary

Acceptable use: A permitted "land use" activity in a designated geographic area, as defined by the municipal "zoning code."

Aeroponics: A method of growing plants in air, without soil. Nutrition is typically provided through an aqueous mist.

Apiary: A collection of honeybee hives, often located in close proximity to crops.

Apiculture: The practice of keeping and tending to honeybee colonies.

Apis mellifera: The honeybee species native to Europe (common name, European honeybee), prized for prolific honey production, rapid population growth and calm temperament, relative to other honey-producing bee species.

Aquaponics: A method of raising aquatic animals and plants in a closed-loop environment, in which animal waste provides nutrients to the plants, and the plants filter waste from the water.

Bare membrane roof: A roof that contains a waterproofing membrane as its uppermost surface. These roofs are typically flat (less than 2% pitch) and lack shingles, vegetation, stone ballast or other protective coverings.

Base irrigation: An irrigation technique that delivers water to plants at the base of the green roof system, through either "drip lines" or a base flood system.

Building code: A set of rules at the local, state, provincial, federal or international level that specify the minimum level of safety for the design and construction of buildings and other structures. The codes are established and enforced to protect public health, safety and welfare.

Building footprint: A projection of the built area of a building's first floor. Certain jurisdictions include overhangs and balconies.

Certified organic: Foods or goods for which the production passed a third-party inspection by a provincial, state or nationally accredited certifying agent recognized by the USDA, Quality Assurance International or other similar entity.

CDC: See Community Design Collaborative.

Chemical-free produce: Fruits, vegetables and herbs grown without the use of synthetic inputs, such as non-organic fertilizer, herbicides and pesticides.

Chicago Rarities Orchard Project (CROP): A Chicago-based organization that promotes urban fruit tree diversity by planting orchards of heritage varieties.

Cistern: A large container, either above or below grade, that holds water. These receptacles can be engineered to capture rainwater and direct the water through an irrigation system or building.

Cold frame: A low-profile transparent structure in which vegetables and herbs are planted. The structure shields crops from the wind and maintains a temperature slightly higher than that outside the frame.

Combined sewer overflow (CSO): A point-source pollution event in which greywater and blackwater are released from sewer pipes or waste treatment plants directly into nearby streams, rivers, lakes or oceans. These events occur when a single sewer system accepts both stormwater and wastewater (typical infrastructure in older American cities), and sewer flow exceeds capacity.

Community Design Collaborative (CDC): A collective of professional designers, engineers and estimators that provide pro bono design work for clients in Philadelphia.

Community Supported Agriculture (CSA): A food distribution technique whereby customers buy a share in a farm's (or a farm collective's) upcoming harvest. In exchange for upfront payment, members receive a weekly or biweekly share of the harvest. Both shareholder and farmer share the season's production risks and benefits.

Companion planting: A method of planting complementary plant species in close proximity to one another. Companion crops, such as corn and beans, benefit each other by increasing nutrient uptake, pest control, pollination, etc.

Compost: Decomposed organic material used as a high-nutrient soil amendment or planting medium. Organic waste called greens, (e.g., vegetable scraps) and bulk agents called browns (e.g., straw, wood chips) are combined in specific ratios to promote microbial activity.

Co-op: An association of volunteers who cooperatively pool their resources in order to achieve a collective economic or social benefit. Food cooperatives typically offer discounts or rebates in exchange for members paying an annual fee and/or working a designated number of hours at the establishment.

CROP: See *Chicago Rarities Orchard Project.*

Crop rotation: The practice of planting crops from a single family in the same location no more than once every three years. This farming technique reduces the occurrence of pests and improves overall crop health.

Crowd funding: The collective process by which individuals pool money and other resources to support an individual or cause.

CSA: *See Community Supported Agriculture (CSA).*

CSO: *See Combined sewer overflow.*

Dairy farming: The class of agriculture in which cows, sheep, goats or other milk-producing animals are kept for the commercial production of milk.

Dead load: A fixed vertical weight, such as mechanical equipment or a roof deck, on a structure. Most green roof professionals classify vegetated roofs as a dead load (rather than a "live load"), but consensus does not exist among structural engineers.

Desiccation: Soil drying, often caused by wind.

DIY: Acronym for Do-it-yourself, a term describing the act of designing, building or repairing something without professional assistance.

Drip lines: Perforated plastic tubing that conducts irrigation water. Lines can be placed above, within or at the base of planting soil.

Economic break-even point: The point at which costs equal revenue, and a company breaks even.

Exposure (to the elements): Susceptibility to detrimental effects from amplified environmental conditions, such as temperature fluctuations and high winds.

Floor area ratio (FAR): Ratio of a building's "gross floor area" to its parcel size.

Food desert: A low-income neighborhood where at least 500 residents, or 33 percent of the population, live more than one mile from a supermarket or large grocery store (in an urban setting) and more than ten miles in a rural setting.

Food localization: Growing and distributing food closer to population centers through home gardening, community gardening, farmers' markets, Community Supported Agriculture (CSA) and retail. Local is generally defined as within 50 miles.

Greenhouse: A glass- or plastic-clad framed structure in which plants are grown. The structure moderates temperature and blocks wind, while retaining heat as the floor absorbs solar radiation. Mechanization and temperature controls are common.

Gross floor area: The combined area of all floors of a building.

Headhouse: A rooftop vestibule in which a staircase exits onto the roof. Building codes often require this structure for rooftop access on residential buildings.

Holistic building design: An architectural approach in which buildings are designed to integrate systems that address water use (and sometimes reuse), indoor temperature, indoor air quality, waste cycling, etc. Outputs of certain building systems are sometimes utilized as inputs of other systems.

Hoop house: A non-temperature-controlled, non-mechanized, plastic-clad structure with Quonset framing. Also known as a "high tunnel."

High tunnel: *See Hoop house.*

Hydroponics: A method of growing plants in a nutrient solution without soil. Systems typically provide plants with an inert substrate in which to root, such as coconut coir, mineral wool, perlite or expanded clay.

Integrated Pest Management (IPM): An environmentally sensitive approach to pest management that reduces pest damage with strategies such as rotating crops, selecting pest-resistant crop varieties, attracting beneficial insects that prey on pests and using non-synthetic pesticides. A common practice in organic agriculture.

Land use: The way in which a geographic area is used, as set forth by the municipal "zoning code." Standard uses include residential, commercial, light manufacturing and open space. Agriculture is sometimes a land use, and other times it is included as an "acceptable use" within a land use.

Line load: The weight, from an object such as a rooftop knee wall, distributed evenly along a linear path. The load is calculated in units of force per unit of length.

Live load: A non-fixed weight, such as a planter or movable partition, on or within a structure. Human occupants and "snow load" are types of live loads, but "building codes" typically specify each load independently, and vary requirements for each according to building use and location. Some structural engineers classify vegetated roofs as a live load, while most green roof professionals consider them a "dead load."

Livestock farming: Class of agriculture in which herd animals are raised for slaughter.

Local: Generally defined as food sold within 50 miles from where it was grown, produced or raised.

Lull: A type of Class 7 reach forklift used commonly in green roof construction on one- to three-story buildings.

Microclimate: An immediate climate that differs from that of surrounding areas in terms of temperature, humidity, wind, solar gain, etc. These areas produce unique growing conditions for plants.

Mixed-use development: A group of buildings in a given area designed for multiple uses simultaneously, such residential, retail and commercial use. These developments are often pursued to reduce daily transportation needs and increase quality of life. "Zoning codes" often specify this as a land-use type.

Midwest Organic Services Association (MOSA): A non-profit agency that evaluates and provides organic certification to farm applicants throughout the US.

Municipal water: Potable water that has been processed by a treatment plant and is available for household, manufacturing and other uses by the municipality.

Open-air hydroponics: A method of growing plants hydroponically, without the protection of a greenhouse or hoop house. The practice is most common in warmer climates, such as Singapore, southern Florida and the tropics.

Orchard production: Class of agriculture in which tree and bush fruit is cultivated.

Organic: *See Certified organic.*

Parapet: Perimeter wall around a flat-roofed building, which can range in height from several inches to many feet.

Particulate: Small grains of wind-borne sediment and dust that accumulate on urban surfaces (e.g., flat roofs), and are filtered from the air naturally by plant leaves.

Party walls: Shared walls between row homes that typically extend as knee-high walls at the roof level, and are more weight bearing than the roof itself.

Peak flow: The highest point of rainfall intensity in inches per hour during a precipitation event. Green roofs decrease peak flow by slowing the rate at which stormwater (*see Stormwater*) exits the roof.

Permaculture: An ecological design system that promotes methods of sustainable food production, building, resource management and living.

Philadelphia Rooftop Farm (PRooF): A group of volunteers and activists that promote rooftop agriculture and build rooftop planter prototypes in Philadelphia.

Photovoltaic panels: An assembly of photovoltaic cells that convert solar radiation to electricity. Also known as PVs or "solar panels."

Point load: Weight, from an object such as a planter or raised bed, concentrated in a single location.

Polyculture: The practice of planting diverse crop species within a given space at a given time to reduce the chances of disease and pests, minimize the need for synthetic herbicides and pesticides, and harvest a wide variety of crops at once. This is a common practice in organic agriculture.

Rain barrel: A plastic or wooden drum designed to collect stormwater runoff from surfaces such as roofs. Typically a gutter segment feeds the barrel at the top, while a spigot releases water at the barrel's bottom.

Raised bed: Built-in-place or prefabricated low-profile structures filled with soil and used to grow plants. Framing is often built of wood, brick, cinder block or metal siding.

Raw: Unpasteurized, in the case of honey or dairy products.

Reservoir sheet: A high-"transmissivity" molded plastic sheet installed at the base of a green roof to retain water for plant uptake while increasing the system's rate of horizontal drainage.

Roofscape: A landscape on-structure, coined by the Philadelphia-based green roof firm Roofscapes, now known as Roofmeadow.

Run rate revenue: The annual revenue of a company when extrapolating the current revenue over a twelve-month period. A useful tool for new companies that are predicting future revenue.

Setback: The no-build zone specified by a local building code. On rooftops, this can apply to the roof zone closest to the street.

Self-watering container: A molded plastic planter or set of stacked plastic bins that contain a false bottom to separate planting soil from a water reservoir at the base of the system. "Separation fabric" often lines the false bottom and prevents soil from entering the reservoir, while allowing plant roots to reach the water. Water is delivered through a watering pipe that connects the surface of the planter to the reservoir.

Separation fabric: A woven geotextile, permeable to water, used to separate materials in a green roof system or line a raised bed. The fabric is available in various densities measured in ounces per square yard (oz/sy) (e.g., 4, 6 or 8oz/sy) with corresponding tensile strengths, puncture resistance strengths and permeability. Also known as landscape fabric.

SHARE Food Program: A non-profit organization based in North Philadelphia that serves a regional network of community organizations engaged in food distribution, education and advocacy. It sells affordable food packages, which are sometimes supplemented with fresh produce from an onsite urban farm.

Sheet drain: A medium-"transmissivity" molded plastic sheet installed at the base of a green roof to retain water for plant uptake while increasing the system's rate of horizontal drainage. The material typically contains separation fabric, adhered to one or both sides of the sheet.

Single-source warranty: A warranty offered by a waterproofing manufacturer that jointly covers the green roof and underlying waterproofing system. The cost of uncovering the overburden (in the rare case of a leak) is paid for by the warranty provider, rather than the building owner.

Snow load: The weight of accumulated snow on a structure. "Building codes" in colder climates specify a building's design snow load, which is calculated in addition to "dead load" and "wind load."

Solar panels: *See Photovoltaic panels.*

Square foot gardening: The agricultural practice of planting intensive garden plots to achieve higher yields than typically achieved in a plot of the same size.

Stormwater: Rainwater or snowmelt that either infiltrates a permeable surface or runs across an impermeable surface (*see Stormwater runoff*).

Stormwater runoff: Rainwater or snowmelt that flows across impermeable surfaces such as asphalt, concrete or bare roof membranes. Runoff may carry pollutants from the surfaces and deposit then into streams, rivers, ocean or the sewer system.

Sub-surface irrigation: An irrigation technique for delivering water to plants below the surface of the soil or media through "drip lines." Lines are typically buried at least two inches.

Super: A wooden box-like structure within a Langstroth beehive that contains 8 to 10 frames to collect honeycomb. These boxes vary in height and are generally stacked.

Tear-off and reroof: A "waterproofing membrane" that is fully removed and replaced.

Transmissivity: The rate at which water flows horizontally through a green roof or aquifer. Green roof drainage layers (e.g., "reservoir sheets," "sheet drains," moisture management mats, granular drainage media) exhibit varying rates of horizontal flow.

Triple bottom line: A business philosophy asserting that a business cannot survive long-term without ensuring social, environmental and economic sustainability. The slogan "people, planet and profit" often applies.

Urban heat island effect: A consistent and measurable temperature differential between a city and the non-urban surrounding geography, generally caused by a lack of vegetation, soil, water and light-colored surfaces in the city.

US Department of Agriculture (USDA): The federal department within the United States government that provides leadership on food, agriculture and natural resources. The USDA develops policy and framework based on efficient management and the best available science.

Value-added products: Minimally processed goods (e.g., jelly, hot sauce, candles) that are sold at a higher price than the raw goods from which they are made.

Vermiculture: The composting technique that utilizes certain types of worms, typically *Eisenia foetida* (Red Wigglers), to break down organic matter.

Victory Garden: The edible home and parkland gardens planted by Americans, Canadians and Europeans during World War I and II. Food from the gardens empowered citizens and alleviated the public food-supply demand while farmers were at war.

Vineyard production: The class of agriculture in which vine fruit is cultivated.

Waterproofing membrane: A material that is applied to roofs to keep water out of the building below. Common materials include polyvinyl chloride (PVC), thermoplastic polyolefin (TPO), rubberized asphalt, modified bitumen and ethylene propylene diene monomer rubber (EPDM). Liquid materials are hot- or cold-applied. Single-ply materials are loose-applied or adhered.

Wind load: The static force that wind exerts on a roof or rooftop structure (e.g., greenhouse). Local "building codes" specify a building's design wind load, which is calculated in addition to "dead load" and "live load."

Wind uplift: A net upward force cause by an air pressure differential above and below a surface, such as a roof or raised bed.

Winnowing: Soil loss caused by wind.

Work party: A gathering of volunteers and staff members (when applicable) that work collectively on a specified day to achieve a predetermined goal.

Zoning code: Municipal law that specifies how and for what purpose each "land use" within a city can be used. This planning tool allows cities to guide future development and protect natural resources. The municipal "zoning map" illustrates the code.

Zoning map: A municipal map that delineates the specified "land use" for each geographic area of a city and corresponds to the municipal "zoning code."

Endnotes

Chapter 1 The Backdrop

1. Geoff Wilson. *Why Urban Rooftop Microfarms Are Needed for Sustainable Australian Cities*. The Urban Agriculture Network, Western Pacific, 2004.
2. Jac Smit. *Urban Agriculture and the 21st Century*. City Farmer, 2007.
3. Michael Ableman. "Agriculture's Next Frontier: How Urban Farms Could Feed the World." *Utne Reader*, no. 102, 2000.
4. Catherine Murphy. "Cultivating Havana: Urban Agriculture and Food Security in the Years of Crisis." *Institute for Food and Development*, no. 12, 1999.
5. Wilson, 2004.

Chapter 2 What's In It for Me?

1. C.L. Ogden, M.D. Carroll, B.K. Kit & K.M. Flegal. *Prevalence of Obesity in the United States, 2009–2010*. US Department of Health and Human Services, Centers for Disease Control and Prevention, National Center for Health Statistics data brief, no. 82, 2012.
2. J. Maas, R.A. Verheij, P. P. Groenewegen, S. de Vries & P. Spreeuwenberg. "Green Space, Urbanity, and Health: How Strong Is the Relation?" *Journal of Epidemiology and Community Health*, Vol. 60, 587-592, 2006.
3. J. Maas, R.A. Verheij, S. de Vries, P. Spreeuwenberg, F.G. Schellevis & P. P. Groenewegen. "Morbidity Is Related to a Green Living Environment." *Journal of Epidemiology and Community Health*, Vol. 63, 967–973, 2009.
4. Sonja M.E. Van Dillen, Sjerp De Vries, Peter P. Groenewegen & Peter Spreeuwenberg. "Greenspace in Urban Neighbourhoods and Residents' Health: Adding Quality to Quantity." *Journal of Epidemiology and Community Health*, Vol. 66, no. 8, 2011.
5. T. Takano, K. Nakamura & M. Watanabe. "Urban Residential Environments and Senior Citizens' Longevity in Megacity Areas: The Importance of Walkable Green Spaces." *Journal of Epidemiology and Community Health*, Vol. 56, 913-918, 2002.

6. Ismael Hautecœur. The Rooftop Garden Project, Project Coordinator. Interview conducted January 6, 2010.

7. Community Food Security Coalition North American Urban Agricultural Committee. *Urban Agriculture and Community Food Security in the United States: Farming from the City Center to the Urban Fringe*, 2003.

8. R. Pirog, T. Van Pelt, K. Enshayan & E. Cook. *Food, Fuel, and Freeways: An Iowa Perspective on How Far Food Travels, Fuel Usage, AND Greenhouse Gas Emissions.* Leopold Center for Sustainable Agriculture, Iowa State University: Ames, IA, 2001.

9, 10. US Department of Agriculture, Economic Research Service. *Food Desert Locator*, 2012. Accessed August 4, 2012. www.ers.usda.gov/data-products/food-desert-locator/documentation.aspx.

Chapter 3 Seed to Plate

1. Gotham Greens. *Our Greenhouse in Brooklyn*, 2012. Accessed August 27, 2012. gothamgreens.com/our-farm.

2. Kurt D. Lynn. *Lufa Farms: A Successful Commercial Rooftop Farm in Montreal.* Urban Agriculture Summit, lecture and panel discussion. Ryerson University, Toronto, ON, August 16, 2012.

3. Gotham Greens.

4. Steven E. Newman, n.d. *Greenhouse Structures.* Colorado State University Cooperative Extension, Horticulture and Landscape Architecture.

5. US Department of Agriculture. *Code of Federal Regulations. Title 7, Part 205*, National Organic Program, 2012. Accessed August 26, 2012. ecfr.gpoaccess.gov.

6. Al Johnson. *A Conversation Between Al Johnson and Norbert Blei.* Door County Icon Series podcast, 2007. Interview conducted February 20, 2007.

7. Hannah Nordhaus. *The Beekeepers Lament: How One Man and Half a Billion Bees Help Feed America.* Harper Perennial, New York, 2010.

8. Trey Flemming. Urban Apiaries, Founder. Interview conducted August 26, 2011.

9. NYC Board of Health. Wild and Other Animals Prohibited. *Health Code Text Amendment: Art. 161.01.* Enacted March 16, 2010.

10. Mohamed Hage. Lufa Farms, Founder and President. Interview conducted August 13, 2012.

Chapter 5 Rooftop Farms [Medium-scale]

1. Annie Novak. Eagle Street Rooftop Farm, Cofounder and Farmer. Interview conducted October 23, 2011.

2. Dave Snyder moved on to other Midwestern agricultural pursuits since recording this interview in 2012. His exceptional work at Uncommon Ground will be long remembered.

3. Dave Snyder. Uncommon Ground, Rooftop Farm Director. Interview conducted May 15, 2012.

CHAPTER 6 ROOFTOP AGRICULTURE INDUSTRY [LARGE-SCALE]

1. Viraj Puri. Gotham Greens, Cofounder and CEO. Interview conducted September 8, 2011.

2. Vincent Dessberg. I Grow My Own Veggies, Founder. Interview conducted January 20, 2012.

3. Dana Hauser. The Fairmont Waterfront, Executive Chef. Interview conducted August 8, 2012.

4. Ben Flanner. Brooklyn Grange, Head Farmer and President. Interview conducted August 20, 2012.

5. Ben Flanner. *The Brooklyn Grange*. Climate, Cities and Behavior Symposium. Garrison Institute, April 28, 2011.

6, 7. Paul Lightfoot. *Scaling up Local Food: BrightFarms and the Business of Urban Agriculture*. Urban Agriculture Summit, keynote address. Ryerson University, August 16, 2012.

8. Lynn, 2012.

9. Hage, 2012.

10. Ben Flanner. *Wearing Many Hats: Growing Food on New York City Roofs*. Urban Agriculture Summit, lecture and panel discussion. Ryerson University, August 16, 2012.

11. Flanner, interview, 2012.

12. NYC Department of Environmental Protection. *DEP Awards $3.8 Million in Grants for Community-Based Green Infrastructure Program Projects*, 2011. Accessed August 22, 2012.

13. Lynn, 2012.

14. United States Cong. Senate, 111th Congress. *S. 320 Sec. 506, Clean Energy Stimulus and Investment Assurance Act of 2009*, 2009. Accessed August 25, 2012. thomas.loc. gov/cgi-bin/query/F?c111:1:./temp/~c111C9myYI:e56190.

15. City of Chicago, Buildings Department. *DOB Green Permit Requirements*, 2012. Accessed August 26, 2012. www.cityofchicago.org/city/en/depts/bldgs/provdrs/green_permit.html.

16. City of Chicago, Department of Streets and Sanitation. Compost Standards for Chapter 11-4 Permit Exempt Compost Facilities Rules and Regulations. *Chicago Municipal Code: Sec. 7-28-710 and 7-28-715*. Enacted June 7, 2007.

17, 18. City of Chicago, Plan Commission. Zoning Code Ordinance Amendment. *Municipal Code of Chicago: Title 17 Sec. 17-3-0207, 17-4-0207, 17-5-0207, and 17-6-0403-F*, 2012.

19. City of Chicago, Plan Commission. Zoning Code Ordinance Amendment. *Municipal Code of Chicago: Title 17 Sec. 17-17-0270.7 and 17-17-0270.6, Title 07 Sec. 07-12-210 and 07-12-300*, 2012.

20. City of Philadelphia, Department of Revenue. Business Privilege Tax: Green Roof Tax Credit. *Bill No. 070072*. Enacted April 12, 2007.

21. City of Philadelphia, Water Department. Press release. *PWD and PIDC Award $3.2 Million in Grants to Promote Green Stormwater Management Practices on Private and Non-Profit Properties Resulting in the Planned Development of 64 Green Acres*, 2012.

22. City of Philadelphia, City Planning Commission. Use-Specific Standards. *Zoning Code: 14-603 Sec. 15*. Effective August 22, 2012.

23. City of Philadelphia, City Planning Commission. Use-Specific Standards. *Zoning Code: 10-100 Sec. 10-101 and 10-112*. Effective August 22, 2012.

24. City of Portland. Floor Area and Height Bonus Options. *Building Code: 33.510.210*. Enacted March 5, 2010.

25. City of Portland, Bureau of Environmental Services. Grey to Green Initiative. *EcoRoof Grants*. Effective July 1, 2008

26, 27. City of Portland, Bureau of Planning and Sustainability. *Urban Food Zoning Code Update Concept Report*, 2011.

28. City of San Francisco. Green Building Ordinance. *San Francisco Building Code: Chapter 13C*. Enacted 2008.

29. City of San Francisco, Board of Supervisors. Urban Agriculture Ordinance. *San Francisco Administrative Code: Chapter 53*. Enacted 17 July 2012.

30. City of San Francisco. Planning Code: Urban Agriculture. *Planning Code Ordinance Amendment: Sec. 102.34, 204.1, 209.5, 227, 234.1, 234.2, and Art.7 and 8*. Enacted 20 April 2011.

31. City of San Francisco, Department of Health. Animal Care and Control. *San Francisco Health Code: Sec. 12, 27, 32, 37*, 2012. Enacted various dates.

32. New York State. Green Roof Property Tax Abatement. *Laws of New York: Chapter 461/4-B*, 2009. Effective 1 January 2009 to 15 March 2013.

33. City of New York, Department of Environmental Protection. *Green Infrastructure*

Grant Program, 2012. Accessed August 19, 2012. www.nyc.gov/html/dep/html/
stormwater/nyc_green_infrastructure_grant_program.shtml.

34. City of New York, Building Code. 2012. Certification for Rooftop Greenhouses.
Zone Green Text Amendment: N 120132 ZRY/75-01. Enacted April 30, 2012.

35. City of New York, City Planning Commission. *Zoning Code: Art. II Sec. 22-14, Art.
III Sec. 32, Art. IV Sec. 42-14.* Enacted February 2, 2011.

36. City of New York, Department of Health and Mental Hygiene. Definitions. *Health
Code Text Amendment: Article 161.02.* Enacted 16 March 2010.

37. City of New York, Department of Health and Mental Hygiene. Permits to keep
certain animals. *Health Code Amendment: Article 161.09.* Enacted 16 March 2010.

38. City of Vancouver. Port Coquitlam Green Roof Bylaw. *Zoning Bylaw Amendment:
No. 2240.* Adopted December 11, 2006.

39. City of Vancouver. *Vancouver Food Charter.* Adopted January 2007.

40. Metro Vancouver. *Metro Vancouver Regional Food System Strategy.* February 2011.

41. City of Vancouver. Urban Agriculture Guidelines for the Private Realm. *Policy Report:
Development and Building: 08-2000-20.* Adopted 20 January 2009.

42. City of Vancouver. *Zoning & Development Bylaw 3575.* Effective January 1, 2012.

43. City of Vancouver. Hobby Beekeeping. *Planning: By-law Administration Bulletins.*
Effective February 27, 2006.

44. City of Vancouver. *Animal Control By-law No. 9150.* Effective May 29, 2012.

45. City of Toronto. *Green Roof By-law No. 538-2009. Municipal Code: 492.* Adopted
May 27, 2009.

46. Greater Toronto Area Agricultural Action Committee, 2012. *Golden Horseshoe Food
and Farming Action Plan 2021.*

47. City of Toronto, 2007. *Zoning By-Law: 100.10, 1.40.90.*

48. City of Toronto. Animals. *Toronto Municipal Code: Chapter 349.* Adopted March 31,
2009.

49. City of Montreal, 2010. Permanent Agricultural Zone. *R.S.Q.: Chapter P-41.1.*

50. City of Montreal. Uses Authorized in All Zones. *Zoning By-Law CA29 0040:
Chapter 4, Sec. 9, Art. 58.* Adopted June 7, 2010.

51. City of Montreal. Uses Authorized in All Zones. *Zoning By-Law CA29 0040:
Chapter 4, Sec. 7.* Adopted June 7, 2010.

52. Ibid.

Chapter 7 Potential Hotspots

1. City of Philadelphia, City Planning Commission. Use-Specific Standards. *Zoning*

Code: 14-603 Sec. 15. Effective August 22, 2012.

2. City of Portland, Bureau of Planning and Sustainability, 2010. *Ecoroofs*. Accessed 20 March 2010. www.portlandonline.com/bps/index.cfm?c=ecbbd&a=bbehci.

3. Bay Localize. Tapping the potential of urban rooftops: Rooftop resources neighborhood assessment. Earth Island Institute, 2007.

4. United States Census Bureau. *State and County QuickFacts: New York (city), New York. 2010 Census.*

5. Statistics Canada. *The Canadian Population in 2011: Population Counts and Growth. Population and dwelling counts, 2011 Census. No. 98-310-X2011001.*

6, 7. Stacey McDonald, Kelsey Norman & Nina Damsbaek. *The Potential of Rooftop Agriculture in Toronto, Canada*. University of Toronto, research paper, 2009.

8. City of Toronto. *Green Roof By-Law No. 538-2009*, 2009.

9. Statistics Canada. 2011.

CHAPTER 8 THE PATH TO MARKET

1. Novak, 2011.

Bibliography

Ableman, Michael. "Agriculture's Next Frontier: How Urban Farms Could Feed the World." *Utne Reader*, no. 102, 2000.

Bay Localize. *Tapping the Potential of Urban Rooftops: Rooftop Resources Neighborhood Assessment*. Earth Island Institute, 2007.

Bosse, Alexa. Philadelphia Rooftop Farm Project (PRooF), Architectural Consultant. Personal communication, March 26, 2010.

City of Chicago, Buildings Department. *DOB Green Permit Requirements*. City of Chicago, 2012. Accessed August 26, 2012. www.cityofchicago.org/city/en/depts/bldgs/provdrs/green_permit.html.

City of Chicago, Department of Streets and Sanitation. Compost Standards for Chapter 11-4 Permit Exempt Compost Facilities Rules and Regulations. *Chicago Municipal Code: Sec. 7-28-710 and 7-28-715*. Enacted June 7, 2007.

City of Chicago, Plan Commission. Zoning Code Ordinance Amendment. *Municipal Code of Chicago: Title 17 Sec. 17-3-0207, 17-4-0207, 17-5-0207, and 17-6-0403-F*. City of Chicago, 2012.

City of Chicago, Plan Commission. Zoning Code Ordinance Amendment. *Municipal Code of Chicago: Title 17 Sec. 17-17-0270.7 and 17-17-0270.6, Title 07 Sec. 07-12-210 and 07-12-300*. City of Chicago, 2012.

City of Philadelphia, City Planning Commission. Use-Specific Standards. *Zoning Code: 10-100 Sec. 10-101 and 10-112*. Effective August 22, 2012.

City of Philadelphia, City Planning Commission. Use-Specific Standards. *Zoning Code: 14-603 Sec. 15*. Effective August 22, 2012.

City of Philadelphia, Department of Revenue. Business Privilege Tax: Green Roof Tax Credit. *Bill No. 070072*. Enacted April 12, 2007.

City of Portland, Bureau of Environmental Services. Grey to Green Initiative. *EcoRoof Grants*. Effective July 1, 2008.

City of Portland, Bureau of Planning and Sustainability. *Ecoroofs*. Accessed March 20, 2010. City of Portland, 2010. www.portlandonline.com/bps/index.cfm?c=ecbbd&a=bbehci.

City of Portland, Bureau of Planning and Sustainability. *Urban Food Zoning Code Update Concept Report*. City of Portland. 2011.

City of Portland. Floor Area and Height Bonus Options. *Building Code: 33.510.210*. Enacted March 5, 2010.

City of Montreal. Permanent Agricultural Zone. *R.S.Q.: Chapter P-41.1*. City of Montreal. 2010.

City of Montreal. Uses Authorized in All Zones. *Zoning By-Law CA29 0040: Chapter 4 Sec. 7*. Adopted June 7, 2010.

City of Montreal. Uses Authorized in All Zones. *Zoning By-Law CA29 0040: Chapter 4 Sec. 9 Art. 58*. Adopted June 7, 2010.

City of Philadelphia, Water Department. *PWD and PIDC Award $3.2 Million in Grants to Promote Green Stormwater Management Practices on Private and Non-Profit Properties Resulting in the Planned Development of 64 Green Acres*. Press release. City of Philadelphia, 2012.

City of San Francisco, Board of Supervisors. Urban Agriculture Ordinance. *San Francisco Administrative Code: Chapter 53*. Enacted July 17, 2012.

City of San Francisco, Department of Health. Animal Care and Control. *San Francisco Health Code: Sec. 12, 27, 32, 37*. Enacted various dates. City of San Francisco. 2012.

City of San Francisco. Green Building Ordinance. *San Francisco Building Code: Chapter 13C*. Enacted 2008.

City of San Francisco. Planning Code: Urban Agriculture. *Planning Code Ordinance Amendment: Sec. 102.34, 204.1, 209.5, 227, 234.1, 234.2, and Art.7 and 8*. Enacted April 20, 2011.

City of Toronto. *Zoning By-Law: 100.10, 1.40.90*. City of Toronto. 2007.

City of Toronto. Animals. *Toronto Municipal Code: Chapter 349*. Adopted March 31, 2009.

City of Toronto. Green Roof By-Law No. 538-2009. *Municipal Code: 492*. Adopted May 27, 2009.

City of Vancouver. Hobby Beekeeping. *Planning: Bylaw Administration Bulletins*. Effective February 27, 2006.

City of Vancouver. Port Coquitlam Green Roof Bylaw. *Zoning Bylaw Amendment: No. 2240*. Adopted December 11, 2006.

City of Vancouver. *Vancouver Food Charter*. Adopted January 2007.

City of Vancouver. Urban Agriculture Guidelines for the Private Realm. *Policy Report: Development and Building: 08-2000-20*. Adopted January 20, 2009.

City of Vancouver. *Zoning & Development Bylaw 3575*. Effective January 1, 2012.

City of Vancouver. *Animal Control By-Law No. 9150*. Effective May 29, 2012.

Community Food Security Coalition North American Urban Agricultural Committee. *Urban Agriculture and Community Food Security in the United States: Farming from the City Center to the Urban Fringe*. Community Food Security Coalition, 2003.

Cooper, Harley. CNT: Sustainable Communities Attainable Results, Geographic Information Systems Analyst. Interview conducted March 18, 2010.

Dessberg, Vincent. I Grow My Own Veggies, Founder. Interview conducted January 20, 2012.

Fairmont Waterfront. *Executive Chef Dana Hauser at the Fairmont Waterfront, Vancouver*. Fairmont Hotels & Resorts, 2012.

Flanner, Ben. *The Brooklyn Grange*. Climate, Cities and Behavior Symposium. Garrison Institute, April 28, 2011.

Flanner, Ben. *Wearing Many Hats: Growing Food on New York City Roofs*. Urban Agriculture Summit, lecture and panel discussion, Ryerson University, August 16, 2012.

Flanner, Ben. Brooklyn Grange, Head Farmer and President. Interview conducted August 20, 2012.

Flemming, Trey. Urban Apiaries, Founder. Interview conducted August 26, 2011.

Friehling, Benji. Kibbutz farmer. Interview conducted January 20, 2010.

Goldberg, Lindsey. Graze the Roof and Glide Foundation, Project Manager and Garden Educator. Interview conducted June 7, 2012.

Goode, Lisa. Goode Green, Principal and Eagle Street Farm Landscape Architect. Interview conducted March 21, 2010.

Gotham Greens. *Our Greenhouse in Brooklyn*. Accessed August 27, 2012. gothamgreens.com/our-farm. Gotham Greens, 2012.

Greater Toronto Area Agricultural Action Committee. *Golden Horseshoe Food and Farming Action Plan 2021*. Greater Toronto Area Agricultural Action Committee, 2012.

Hage, Mohamed. Lufa Farms, Founder and President. Interview conducted August 13, 2012.

Hauser, Dana. The Fairmont Waterfront, Executive Chef. Interview conducted August 8, 2012.

Hautecœur, Ismael. The Rooftop Garden Project, Project Coordinator. Interview conducted January 6, 2010.

Johnson, Al. *A Conversation Between Al Johnson and Norbert Blei*. Door County Icon Series podcast. Interview conducted February 20, 2007.

Kortright, Robin. *Evaluating the potential of green roof agriculture*. Trent University NSERC funded project, 2001.

Lightfoot, Paul. *Scaling up Local Food: BrightFarms and the Business of Urban Agriculture*. Urban Agriculture Summit, keynote address. Ryerson University, August 16, 2012.

Lynn, Kurt D. *Lufa Farms: A Successful Commercial Rooftop Farm in Montreal*. Urban Agriculture Summit, lecture and panel discussion. Ryerson University, August 16, 2012.

Maas, J., Verheij, R.A., de Vries, S., Spreeuwenberg, P., Schellevis, F.G. and Groenewegen, P.P. "Morbidity Is Related to a Green Living Environment." *Journal of Epidemiology and Community Health*, Vol. 63, 967–973, 2009.

Maas, J., Verheij, R.A., Groenewegen, P.P., de Vries, S. and Spreeuwenberg, P. "Green Space, Urbanity, and Health: How Strong Is the Relation?" *Journal of Epidemiology and Community Health*, Vol. 60, 587-592, 2006.

McDonald, S., Norman, K. and Damsbaek, N. *The Potential of Rooftop Agriculture in Toronto, Canada*. University of Toronto, research paper, 2009.

Metro Vancouver. *Metro Vancouver Regional Food System Strategy*. February 2011. Metro Vancouver, 2012.

Miller, Charlie. *Vegetated Roof Covers: A New Method for Controlling Runoff in Urbanized Areas*. Pennsylvania Stormwater Management Symposium at Villanova University, October 21-22, 1998.

Miller, Charlie. Roofmeadow, President. Interview conducted September 22, 2012.

Mitchell, Luke. University of Pennsylvania, Masters of City Planning Candidate. Personal correspondence, April 2, 2010.

Mukherjee, Anita. University of Pennsylvania, Wharton School, Applied Economics PhD Candidate. Personal correspondence, April 28, 2010.

Murphy, Catherine. "Cultivating Havana: Urban Agriculture and Food Security in the Years of Crisis. Institute for Food and Development, no. 12, 1999.

New York City. Building Code. Certification for Rooftop Greenhouses. *Zone Green Text Amendment: N 120132 ZRY/75-01*. Enacted April 30, 2012.

New York City, Board of Health. Wild and Other Animals Prohibited. *Health Code Text Amendment: Art. 161.01*. Enacted March 16, 2010.

New York City, City Planning Commission. *Zoning Code: Art. II Sec. 22-14, Art. III Sec. 32, Art. IV Sec. 42-14*. Enacted February 2, 2011.

New York City, Department of Environmental Protection. *Green Infrastructure Grant Program*. Accessed August 19, 2012. www.nyc.gov/html/dep/html/stormwater/nyc_green_infrastructure_grant_program.shtml.

New York City, Department of Environmental Protection. *DEP Awards $3.8 Million in Grants for Community-Based Green Infrastructure Program Projects*. NYC, 2011. Accessed August 22, 2012.

New York City, Department of Health and Mental Hygiene. Definitions. *Health Code Text Amendment: Article 161.02*. Enacted March 16, 2010.

New York City, Department of Health and Mental Hygiene. Permits to keep certain animals. *Health Code Amendment: Article 161.09*. Enacted March 16, 2010.

New York State. Green Roof Property Tax Abatement. *Laws of New York: Chapter 461/4-B*. Effective 1 January 2009 to 15 March 2013.

Nordhaus, Hannah. *The Beekeepers Lament: How One Man and Half a Billion Bees Help Feed America*. Harper Perennial, 2010.

Novak, Annie. Eagle Street Rooftop Farm, Co-Founder and Farmer. Interview conducted October 23, 2011.

Ogden, C.L., Carroll, M.D., Kit, B.K. and Flegal, K.M. *Prevalence of obesity in the United States, 2009–2010*. U.S. Department of Health and Human Services, Centers for Disease Control and Prevention. National Center for Health Statistics Data Brief, no. 82, 2012.

Pirog, R., Van Pelt, T., Enshayan, K. and Cook, E. *Food, Fuel, and Freeways: An Iowa Perspective on How Far Food Travels, Fuel Usage, and Greenhouse Gas Emissions.* Leopold Center for Sustainable Agriculture, Iowa State University, 2001.

Puri, Viraj. Gotham Greens, Co-Founder and CEO. Interview conducted September 8, 2011.

Quinn, James and Trinklein, David. *Master Gardener Core Manual: Vegetable Gardening.* University of Missouri Extension, Division of Plant Sciences, 2008.

Riggall, Gavin. North Street Design, Partner and CDC Team Member. Interview conducted June 24, 2012.

Sand, Jay. Philadelphia Rooftop Farm (PRooF), Organizer. Interviews conducted February 18, and June 16, 2012.

Schwinger, Clifford W. The Harman Group Structural Engineers, Vice President. Personal communication, February 21, 2010.

Sen, Aditi. University of Pennsylvania, Wharton School, Health Economics PhD Candidate. Personal correspondence, April 27, 2010.

Smit, Jac. *Urban Agriculture and the 21st Century.* City Farmer, 2007.

Snyder, Dave. Uncommon Ground, Rooftop Farm Director. Interview conducted May 15, 2012.

Statistics Canada. *The Canadian Population in 2011: Population Counts and Growth.* Population and Dwelling Counts, 2011 Census. No. 98-310-X2011001. Statistics Canada, 2011.

Takano, T., Nakamura, K. and Watanabe, M. 2002. "Urban Residential Environments and Senior Citizens' Longevity in Megacity Areas: The Importance of Walkable Green Spaces." *Journal of Epidemiology and Community Health*, Vol. 56, 913-918.

Templeton, Karen. Interviewed by Marty Logan and Mark Foss. *Urban Agriculture Reaches New Heights Through Rooftop Gardening.* The International Research Development Centre, 2004.

Traunfeld, Jon. *Container Vegetable Gardening: Healthy Harvests from Small Spaces.* Maryland Cooperative Extension, Home and Garden, 2006.

United States Census Bureau. *State and County QuickFacts: New York (city), New York, 2010 Census.* US Census Bureau, 2012.

United States Cong. Senate, 111th Congress. *S. 320 Sec. 506, Clean Energy Stimulus and*

Investment Assurance Act of 2009. US Cong. Senate, 2009. Accessed August 25, 2012. www.thomas.loc.gov/cgi-bin/query/F?c111:1:./temp/~c111C9my YI:e56190.

United States Department of Agriculture. Code of Federal Regulations. *Title 7 Part 205, National Organic Program*, 2012. Accessed 26 August 2012. ecfr.gpoaccess.gov.

United States Department of Agriculture, Economic Research Service. *Food Desert Locator*, 2012. Accessed August 4, 2012. www.ers.usda.gov/data-products/food-desert-locator/documentation.aspx

Valiquette, Marc-André. Biotop Canada Inc., President. Interview conducted January 7, 2010.

Van D., Sonja M.E., de Vries, S., Groenewegen, P.P. and Spreeuwenberg, P. "Greenspace in Urban Neighbourhoods and Residents' Health: Adding Quality to Quantity." *Journal of Epidemiology and Community Health*, Vol. 66, ed. 8, 2011.

Van Hook, Sue. Ecovative Design, Mycological Consultant. Interview conducted April 12, 2010.

Vitiello, Domenic and Nairn, Michael. *Community Gardening in Philadelphia: 2008 Harvest Report*. Philadelphia Harvest, 2009.

Vitiello, Domenic. University of Pennsylvania, Department of City and Regional Planning, Assistant Professor. Personal communication, February 5, 2010.

Warwick, Hugh. "Cuba's Organic Revolution." *The Ecologist*, Vol. 29, no. 8, 1999.

Whiting, David. *Growing Vegetables in a Hobby Greenhouse*. Colorado State University, Cooperative Extension, 2003.

Wikins, Jennifer. *Farm to School in the Northeast: Making the Connection for Healthy Kids and Healthy Farms*. Cornell University, Cornell Farm to School Program, 2007.

Wilkinson, James. *Community Supported Agriculture*. US Department of Agriculture, CD Technotes, 2001. Accessed February 7, 2010. www.rurdev.usda.gov/rbs/CDP-TN20.PDF

Wilson, Geoff. *Why Urban Rooftop Microfarms Are Needed for Sustainable Australian Cities*. The Urban Agriculture Network, Western Pacific, 2004.

Index

About the Author

Lauren Mandel is a Project Manager and Rooftop Agriculture Specialist at Roofmeadow — the preeminent green roof firm in North America — where she designs vegetated and agricultural roof systems and oversees green roof construction throughout the United States. She has visited and photographed rooftop farms and gardens across North America and interviewed prominent rooftop farmers, CEOs, and designers. Ms. Mandel is a contributing writer for *Urban Farm* and *Grid* magazines, a guest lecturer at universities and conferences, and a blogger at EatUpAg.wordpress.com. Ms. Mandel holds a Master of Landscape Architecture from the University of Pennsylvania and a Bachelor of Arts in Environmental Science from Skidmore College. She has previously worked as a landscape designer, US Forest Service wilderness ranger, organic farm intern, and a research intern for American Farmland Trust.

GEOFFREY GOLDBERG PHOTOGRAPHY

If you have enjoyed *Eat Up,* you might also enjoy other

BOOKS TO BUILD A NEW SOCIETY

Our books provide positive solutions for people who want to
make a difference. We specialize in:

**Sustainable Living • Green Building • Peak Oil • Renewable Energy
Environment & Economy • Natural Building & Appropriate Technology
Progressive Leadership • Resistance and Community
Educational & Parenting Resources**

New Society Publishers

ENVIRONMENTAL BENEFITS STATEMENT

New Society Publishers has chosen to produce this book on recycled paper made with
100% post consumer waste, processed chlorine free, and old growth free.

For every 5,000 books printed, New Society saves the following resources:[1]

36	Trees
3,299	Pounds of Solid Waste
3,630	Gallons of Water
4,735	Kilowatt Hours of Electricity
5,997	Pounds of Greenhouse Gases
26	Pounds of HAPs, VOCs, and AOX Combined
9	Cubic Yards of Landfill Space

[1]Environmental benefits are calculated based on research done by the Environmental Defense Fund and
other members of the Paper Task Force who study the environmental impacts of the paper industry.

For a full list of NSP's titles, please call 1-800-567-6772 *or check out our website* at:
www.newsociety.com